# Handbook of Industrial Lighting

# Handbook of Industrial Lighting

**Stanley L. Lyons, FCIBS**

**Butterworths**
London Boston Sydney Wellington Durban Toronto

First published 1981



**British Library Cataloguing in Publication Data**
Lyons, Stanley L.
Handbook of industrial lighting.
1. Industrial buildings — Great Britain — Lighting
2. Electric lighting — Great Britain
I. Title
621.32′2 TK4399.F2

ISBN 0-408-00525-4

Typeset by Tunbridge Wells Typesetting Services
Printed and bound in England

# Foreword

This is a practical handbook to aid the reader who seeks to achieve good industrial lighting. It is intended for engineers and technologists such as lighting engineers, building services engineers, electrical designers and installers, works engineers and architects. It will also provide important back-up reading for students of these professions.

Designed as a work of reference rather than a textbook, it contains much information not available elsewhere except as articles, pamphlets and papers read before institutions. The information is fully up-to-date, and incorporates many practical ideas developed by the author during his long career in illuminating engineering.

The contents of the *Handbook* are relevant to all applications of lighting for industrial premises, including general lighting, task lighting, lighting for many specific engineering and manufacturing processes, lighting for inspection etc. It deals with the practical steps to be taken to design lighting suited to the environments met in various industries, including food manufacture, papermaking, leather and shoe industry, metal manufacturing trades, foundries and drop-forges, printing, brewing, distilling and soft drinks manufacture etc, dealing with the lighting needed to satisfy the visual tasks in each. It also guides the reader on suiting the lighting equipment to hostile conditions such as damp and steam; flammable dusts, vapours and gases; conductive dusts, corrosive atmospheres; vibration of the structure; soiled and smokey atmospheres, etc.

Information is included on the needs for daylighting industrial buildings, relating this to problems of heat-gain and heat-conservation. It deals with the heat-gain from lighting installations and sets out the elements of integrated environmental design for factories, including controlling heat-flow by structural and architectural design relating to the lighting.

Throughout the text there are many references to good safety practices and the role that factory lighting should play in reducing the frequency of industrial accidents. Emphasis is also placed on quality assurance, and an extensive survey of modern inspection techniques is provided. The contents are presented in short numbered paragraphs, extensively cross-referred for convenience in studying any topic, and a full index to the contents is provided.

# Preface

The subject of lighting for factories and industrial plants appears to be simple, but many companies have lighting installations which are the least effective and the most neglected of their building services. Through lack of knowledge, opportunities are lost to light factories well, to make them more productive, more profitable, safer, and altogether more humane and pleasant workplaces. Much of the published information about lighting is of an educational or academic nature that does not suit the needs of the person who has to make important decisions about lighting immediately; further, as may be expected, the technical literature available from lighting manufacturers tends to put forward the types of equipment that it is their business to sell. Thus there is a need for this *Handbook,* which is offered as a source of guidance in the practical business of specifying, designing, ordering, installing, operating and maintaining lighting for all kinds of industrial premises.

I have long regretted the passing of the old *Lighting Service Bureau* (LSB) in Savoy Hill, London, which was operated by the former Electric Lamp Manufacturers' Association from pre-war days up to 1958. Also, the passing of its successor, the *British Lighting Council* which did a similar job of informing the public about lighting from 1958 to 1968. After 1968, The Electricity Council, in collaboration with organisations in the lighting industry, made excellent efforts to provide reliable information in the form of its many publications; but, sadly, its educational and information work has now been much curtailed. It was in harmony with the work of The Electricity Council that, in 1972, I produced the first edition of my book *Management Guide to Modern Industrial Lighting* of which this book may be regarded as the 'engineer's version'. An updated edition of that work will be published by Butterworths in 1982.

In preparing the present *Handbook,* I have tried to keep it simple, and have glanced back to the publications that were available in past years, and have incorporated much of value that I found there. Certain basic principles do not change, and so I have adopted, adapted and improved on what was written in the past, bringing all the facts up to date and in line with modern technology; and I have peppered the pages with practical examples and helpful hints that have arisen from practical work in lighting design and consultancy. Apart from books on lighting, my researches included collecting together many leaflets, booklets and pamphlets from the LSB, the BLC and The Electricity Council, as well as from other sources—these papers together amounting to a pile nearly 600 mm high. As most of these are now out of print, I have not listed them as references. However, some readers will be glad to know that an even greater collection of historical references to lighting are held in the Science Museum, London, under the title of the *The A. D. S. Atkinson Memorial Collection of Lighting Papers* which I was instrumental in having adopted as a National Archive through the good offices of the Lighting Industry Federation and the Company of Lightmongers.

Although I list a considerable bibliography, it was not possible for me to assign individual books as sources of particular items of information, for so much of the data presented in these pages is coloured by my own experiences in lighting. The art and science of illumination does not stand still; up to the last day before despatching my manuscript to the publishers, I have added the latest information culled from lighting manufacturers, researchers and academics. As far as I could contrive it, the information is up-to-date, complete, and incorporates proven guidance on all salient aspects of the specification, design, purchasing, installation, operation and maintenance of good lighting for industry. Importantly, the guide-lines I give are not biassed by any commercial considerations of my own—for I have none—but are intended to give a balanced view of the equipments and methods to be employed to produce the right quantity of light, light of the right quality, and light which has the lowest cost-in-use, while making wise use of both capital and energy.

One of the important reasons for producing a book of this kind is to attempt to bridge the communication gap between lighting specialists and what C. Dykes Brown has dubbed the 'lighting providers', as well as to make the art and science of illumination a meaningful subject to the users. Lighting engineers use many specialist words which have little or no meaning to the layman. Richard Forster(6) has remarked that "while research has been active, the application (of new lighting techniques) has been almost non-existent . . . Has the lighting industry failed to understand its own jargon?" Throughout this book, I have attempted to give the reader a broad understanding of what good lighting is, and how to achieve it. It has not been my intention to attempt to supplant established references such as the *CIBS/IES Code*(5), the *Technical Reports* and *Guides* of the CIBS, and *Interior Lighting Design*(7), to all of which the reader is frequently referred, but to explain and simplify the ideas which underlie the art and science of illuminating engineering.

I acknowledge the assistance I have had from the publications of The Electricity Council and the Chartered Institution of Building Services, as well as the considerable help from Members of the Lighting Industry Federation who have kindly provided product data.

Stanley L. Lyons

# Contents

Chapter 1

# Benefits of good industrial lighting

One important theme of this book is that the process of vision is aided by the provision of good lighting, and that this will affect worker performance. Generally, improvements in lighting produce improvements in quantity or quality of output, or both (1.1). Within certain limits, this 'productivity effect' can be shown to be related to the quantity of illumination, as shown by tests and measurements made in factories (1.2). Many studies have shown that the frequency and severity of industrial accidents can be significantly reduced by the provision of good lighting—this being confirmed by the extensive experience of users of good lighting (1.3). As may be expected, these benefits tend to offset the cost of the lighting, and it is commonly found that the quantifiable benefits due to improved output or quality of work and the reduction in accidents is many times greater than the cost of the lighting (1.4) as well as contributing to the wellbeing of the staff which results in even greater savings due to reduction in absenteeism and other management problems.

## 1.1 Visual performance affected by lighting

**1.1.1** When a person is working, at least 80 percent of the sensory data necessary for the performance of his task is obtained visually. The *visual performance* of the worker is affected by both the quantity and the quality of the illuminance that reveals the task and its surroundings (1.1.2). Visual performance is the achievement of the worker in the performance of his visual task, which may consist of observing small detail (e.g., reading, using instruments, gauges etc), as well as monitoring his immediate environment (e.g., observing for danger, movement of persons, spatial relationships). In some tasks he will require to judge speeds and distances. The performance of a *visual task* (which is usually taken to mean seeing the details of an object quickly and accurately, so that it can be comprehended) will depend in part on the standard of his vision, and on the available illumination which enables his eyes to attain the necessary level of *visual acuity*.

**1.1.2** Visual task acuity[1] is the capacity of the eye for discriminating between details or objects which are close together, or discriminating the size of a very small object. This capacity is expressed as:

$$\frac{1}{S} = \text{Acuity}$$

where $S$ is the angle of separation in minutes of arc between two lines or points which are just separable by the eye (3.1.2).
The visual acuity of a subject is not constant, but varies from instant to instant according to the available illuminance. For example, one cannot read small print by moonlight, simply because one cannot generate sufficient acuity for that task in that illuminance. Thus, to a greater or lesser degree, any visual task could be handicapped if the subject is not provided with sufficient illuminance to achieve the visual acuity required to resolve the details of that task. If the size of detail to be seen is large, then a lesser illuminance will be required, and conversely; here it must be remembered that when we speak of the 'size of detail', what is meant is the 'apparent size'—viz, the combination of physical dimension and distance from the eye—for it is the angle subtended at the eye by the smallest detail that determines the acuity demand.

**1.1.3** Our definition of acuity (1.1.2) must be modified in practical cases to take account of the effect of contrast, e.g., the reflection factors of the parts of the task to be seen, and the contrast ratio between them and between the immediate background. If these reflectances are low, or if the contrast ratio is low, then a greater illuminance will be required to achieve a level of visual task performance compared with a task in which, though having the same 'apparent size' to be picked out, has higher reflectances and, particularly, higher contrast ratios. It will often be the duty of the lighting engineer to advise that the details of a visual task be modified to enhance the contrast ratio, e.g., by introducing colour contrasts, or by arranging that critical objects may be seen in silhouette, or seen against a darker or less well illuminated background (3.2.6).

**1.1.4** The visual acuity of a subject not only varies with the incident illuminance (1.1.2) but declines throughout his life. Older subjects need greater illuminance than younger subjects to achieve comparable visual task performance. The recommendations for illuminances to be provided (Appendix II) take this age factor into account, and provide for illuminances which will satisfy the needs of persons of normal working age. However, it would be sound practice to specify a somewhat greater illuminance than the standard recommendation in areas which are largely occupied by persons over the age of 40.

**1.1.5** Our eyes evolved over millions of years in light which came from the sun, and, biologically speaking, light which flows from a direction other than from above is an experience for which our instincts and reflexes may not always be prepared. Light which comes to the task from some unusual angle will throw shadows, creating modelling and highlights which may be unusual, resulting in mistakes of perception. For example, light from below may cause the illusion of depressed areas of a surface seeming to be raised, and vice versa (Figure 1.1).

**1.1.6** For electric lighting to illuminate an object so that its colours are recognisable and reasonably faithful to actuality, the spectral power of the light in all parts of the spectrum needs to approximate to that in daylight. The subject will be discussed in greater detail in Chapter 4, and it will suffice here to say that if the spectrum of the incident light is discontinuous (e.g., it contains little or no spectral energy in some bands of the visible spectrum) there will be distortion, or even absence, of colour perception. For most industrial tasks, some deficiency in *colour-rendering* (4.2.3) of the lamps will

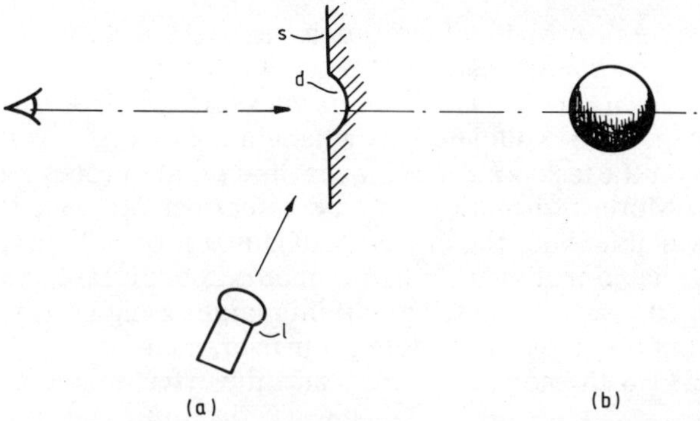


*Figure 1.1* Effects of directional light. (a) Light coming from below eye-level to illuminate a light-coloured depression (d) in a flat plane (s) of lesser reflection factor, may create the illusion that the area (d) is raised rather than depressed. (b) The general appearance of the depression under these lighting conditions is similar to the familiar light pattern seen on a sphere illuminated from above.

not be a serious handicap to efficient work, but we shall also see that reasonable *colour-appearance* (4.2.2) of the lighting is of importance. Thus, lighting that is deficient in colour-rendering property, or is of an unfamiliar and unacceptable colour-appearance, may affect the visual performance of workers, and, directly or indirectly, their task performance (1.2).

**1.1.7** Light which comes to the eye, directly or reflected from objects, which embarrasses vision and handicaps the performance of the visual task is termed *glare.* It is convenient to discuss glare as *discomfort glare* (which does not, at least in the short term, affect the performance of visual tasks, but which tends to bring about an earlier onset of fatigue) and *disability glare* (which handicaps the subject, reducing what he can see—in an extreme case to just seeing the glare-source). Glare may be *direct* (e.g., from unshielded lamps) or *indirect* (e.g., from glossy surfaces or the surface of a liquid). Some surfaces which appear to be *matt,* behave in a *specular* (mirror-like) manner when the light strikes them at low incident angles. Unwanted reflections from the surface of paper, for example, will reduce the *contrast rendering factor* of the print against the paper, to the point where reading is more difficult or is impossible.

**1.1.8** This section has reviewed some of the ways in which the quantity and quality of lighting may affect visual performance. We have seen that visual acuity (1.1.2) is affected by the incident illuminance, by the reflectances and luminance contrasts in the task zone (1.1.3), by the decline of acuity of the subject with age (1.1.4), by unusual directions of light flow (1.1.5), by the colour-rendering property of the light (1.1.6), and by the degree of glare experienced by the subject (1.1.7). These factors are probably the dominant ones, but the list is not exhaustive. Other factors may affect visual performance, such as (a) the time available for seeing, or the speed of movement of the object to be seen; (b) the effect of light which has an interrupted or flickering character; lighting of varying illuminance and its periodicity of variation (5.3); (c) the duration for which the subject is required to perform the task (where, for example, a task can be performed reasonably well for a short period in the given illuminance, but where continuance of the task in that illuminance would lead to early onset of fatigue and incidence of errors). A further factor (d) is the effect of the visual abilities of the subject; for example, a person of normal or corrected vision might perform well under given conditions, while a person suffering from the common eye defect of astigmatism would need a higher illuminance for equal performance, particularly if the task involved frequent and rapid re-focusing of the eyes at distances. More information on how visual performance is affected by lighting is available[1] but the objective of this section will have been reached if it is appreciated that visual performance is not constant and cannot be taken for granted. The art and science of illuminating engineering is directed to providing lighting that will produce a high standard of visual task performance, coupled with economy, safety and the effective use of energy and resources.

## 1.2 Lighting and productivity

**1.2.1** Productivity is the beneficial result of applying resources (manpower, machines, materials, energy, capital, plant and buildings) to the achievement of an objective. The prime activity of management in organizations is (or should be) to maintain and improve the productivity of their organizations. But mere improvement in output may not be enough, for such an improvement must be qualified by stating that increased output at higher unit cost, or increased output at higher reject rate or at lower quality standard, would not be regarded as increases in productivity. Nor would increase in output which was achieved by the subjection of the workforce to greater discomforts or greater risks of injury or ill-health be acceptable. Leaving aside for the moment the humanitarian objections to the latter, and disregarding the objections which would be made by Trade Unions and other organizations should risks of injury or ill-health be imposed on a workforce in quest of greater productivity, it is clear that, although ‘danger money’ might be paid, it is simply not practicable to seek greater productivity at a cost of suffering to others. It is apparent that the simple definition of productivity with which this paragraph opens is not indicative of a policy to be pursued unless the objective of ‘productivity’ is qualified by constraints such as ‘without uneconomic increase in unit cost’, ‘without an uneconomic increase in reject rate or an unacceptable lowering of quality’, and ‘without imposing greater risks or discomforts upon the workforce’. Productivity achieved within all these constraints would almost certainly lead to greater profits by the organization, or the reduction of losses, or the reduction of operating cost of service industries. The objective of this section will be to show that good industrial lighting can aid the achievement of such true productivity, while the next section (1.3) will examine the subject of worker welfare and safety in relation to lighting.

**1.2.2** Studies of work performance have shown that where work is performed in adverse lighting conditions, output and quality of work are lower than may be achieved by the same workforce in optimum lighting conditions. It is well known that the disadvantage of adverse lighting conditions can be partially or temporarily overcome by dint of especial effort by the workers, but that this cannot be long maintained because of stress and the early onset of fatigue. The practice of granting short rest periods has been used as a substitute for improving the lighting, and this has led to bargains in which management agreed to pay the operatives 'relaxation allowances', eg sums of money in lieu of the rest periods. Thus, in some factories, there is the paradoxical situation where the management pays operatives extra wages to work in a poor light! Nothing can be said in favour of this, for the total payments to the workers far exceed what improvement of the lighting would cost (in terms of capital expenditure and running cost), while, even though extra money is paid for the operatives to tolerate the poor lighting, output and quality, and hence profitability, must inevitably suffer.

**1.2.3** Taking an opposite view to those managers who pay their workers 'relaxation allowances' to work in poor lighting (1.2.2), more enlightened managements provide better lighting in the justified expectation that the value of extra goods produced, or the enhanced added-value due to better quality of work, will be greater than the additional cost of providing lighting of optimum standard (18.1). While it must be stressed that the optimum standard of lighting means lighting of good quality as well as of sufficient quantity, the pay-back for this investment and management care may provide increased output, decreased rejects and improvement in safety standards, as indicated in Table 1.

**Table 1** Effect on performance, rejects and accidents due to improvement in illumination in workplaces[2] (See 1.2.3)

| *Company or type of work* | *Illuminance Old (lux)* | *New (lux)* | *Performance increase (%)* | *Rejects decrease (%)* | *Accidents decrease (%)* |
|---|---|---|---|---|---|
| Adox Camera Works | 370 | 1000 | 7.4 | | |
| Mosbach, Gruber & Co Leather punching | 350 | 1000 | 7.6 | | |
| Pearl sorting | 100 | 1000 | 6.0 | | |
| Crocheting | 100 | 1000 | 8.0 | 22 | |
| Classroom test (observation, logical thinking, speedy and accurate calculation) | 90 | 500 | 7.7 (ave.) | | |
| Composing room (print) | 100 | 1000 | 30 | 18 | |
| Screw sorting | 100 | 1000 | 10 | 22 | |
| Linde Machine Factory | 200 | 550 | | | 43 |
| Telephone receiver assy | 150 | 250 | 36 | 57 | |
| Mfr of bearings | 60 | 250 | 13 | | |
| Tile making | 50 | 200 | 4 | | |
| Frawley Corp (Paper Mate Pens) | 500 | 1500 | 28 | | |
| Metwood Mfg Co | 300 | 2000 | 16 | 29 | 52 reduction in lost time |
| Erickson Tool Co | 500 | 1600 2500 | 10 | 20 | 50 of minor accidents |
| Douglas Aircraft (minute parts assy) | 500 general 1000 local | 4000 general | | 90 | |
| Cotton-spinning mill, (loom efficiency) | 170 | 750 | 10.5 loom efficiency | 39.6 | |

**1.2.4** An important function of lighting in factories is to contribute to creating a pleasant environment for the support of morale and for the general wellbeing of those who work there. The lightmeter does not tell the whole story, for while the illuminance on the working plane may be adequate for seeing, the measurement of illuminance does not give any appreciation of how *bright* the room may seem. A room lighted mainly by 'downlighters' for example, especially if the decor were dark, could present a very gloomy and depressing appearance, even if the horizontal illuminance was a 1000 lux or so. The brightness of the room as perceived by the occupants will be due to a combination of the room reflectances and the way the light is distributed from the luminaires and interreflected between the room surfaces (2.2.3).

## 1.3 Lighting and industrial safety

**1.3.1** Of the various factory services which may contribute to the welfare and safety of occupants of workplaces, lighting is of high importance. A report of a five-year study by the Accident Prevention Advisory Unit of the Health and Safety Executive (UK) says that the most safety-conscious companies are those which tend to be commercially successful[(3)]. The report also says that the management characteristics needed to achieve a high standard of health and safety for employees are the same as those required for efficient production. One of the reasons for the failure of some companies to take effective measures for safety is the lack of appreciation and the lack of involvement on the part of senior executives. Decisions about health and safety should rank equal in importance with others regarding the operation of the business. It may well be that many managers do not realise that there is a close link between the standard of lighting provided and the standard of safety to be achieved in industrial premises; thus the responsibility for lighting is often placed with an executive of limited budgetary and decision powers within the organization[(4)]. The director or manager responsible for lighting should acquire a suitable appreciation of the function of lighting in respect of welfare and accident prevention, and should also realise that investment in good lighting is usually self-financing (1.4, 18.2).
**1.3.2** The provision of lighting, 'sufficient and suitable' for the environment and the tasks in the workplace, and in places to which visitors or the public may be admitted, is a legal requirement (see Appendix I). An appreciation of the technical aspects of lighting to enhance safety is given in Chapter 5.

## 1.4 Cost-benefit of good industrial lighting

**1.4.1** Not all the benefits of good lighting can be assessed in financial terms, though there is little doubt that in the majority of cases improvements from poor or mediocre standards to modern standards of lighting will bring measurable returns which can be convincingly demonstrated by calculation. Some of the benefits are due to the improvement of the working environment, and are demonstrated by the modified behaviour of the occupants. For example, improvements in lighting which result in a more pleasant workplace may be accompanied by improvements in *sickness absence rate* (1.4.2), *labour turnover rate* (1.4.3), and *'failed-to-start syndrome' rate* (1.4.4); similarly, losses due to *'invisible absenteeism'* (1.4.5) and *minor accident rate* (1.4.6) are checked. The more tangible benefits, such as reduced errors and faults, greater output, better quality and less accidental damage to goods etc are readily calculated (18.1), as will be the cash savings due to using more efficient luminaires and more efficacious lamps (18.2).

**1.4.2** It has to be recognized that there is widespread abuse of the custom of requiring an employee to produce a doctor's certificate only when a period of absence from work exceeds two days. Keeping a record of the annual lost days due to uncertificated sickness, department by department, or by functions of employees, will prove to be a most interesting indicator of staff satisfaction with the total working environment. Often, when the working environment is significantly improved (e.g. by redecorating, reducing noise-levels or improving the lighting) the loss of productivity and profit due to uncertificated sick-absence will be found to be greatly reduced, and permanently so.

**1.4.3** Labour turnover is, of course, related to wages and other conditions; but, the effect of lighting on the morale of the workforce is usually sufficiently marked for it to be noted that the labour turnover (e.g. rate of replacement of staff per annum due to resignations other than retirement) is lower in better-lighted premises. The difference may show up between departments, or between parallel establishments of the same organization, or between organizations. In companies that provide really good environmental conditions, there may actually be a waiting list of persons who want to join the company—many of whom are relatives, friends or neighbours of existing staff who have heard how good the conditions are. Even in times of mass unemployment, there is always a shortage nationally and locally of really desirable employees; and the provision of good working conditions, including good lighting, is a relatively inexpensive aid to attracting the right people.

**1.4.4** The 'failed to start syndrome' is the situation where a person is hired, but does not turn up for work on the agreed day. It has been noted that if the prospective employee is shown the workplace at the time of the interview and the appearance of the workplace is unpleasant, dingey and poorly lit, then there is a greater probability of the new person not starting work. If the candidates are selected and engaged, but are not shown the workplace at the time of interview, and if the workplace is of unsatisfactory appearance and lighting, then a high proportion of starters resign after a short period of employment.

**1.4.5** 'Invisible absenteeism' is the condition where an employee is on the premises during working hours, but is absent from his workplace for significantly long periods of the day, or is at his workplace but not actually working. In many factories it has been found that a large proportion of female operatives doing repetitive and rather boring jobs may be 'invisibly absent' for as much as 20 minutes in each hour. Improving the lighting may reduce the strain on employees, and make the workplace more bearable, so that improvements of as much as an extra ten minutes in the hour have been noted, ie, output hours raised from 0.66 of norm to 0.83, an improvement of 17% for the relatively insignificant cost of improving the lighting. This effect appears to come about because improving the lighting improves the operatives' distant vision, thus they feel less isolated from the others. The feeling of isolation may be particularly strong in departments where the noise level prevents ordinary conversation between persons at their normal work positions[(4)].

**1.4.6** It is the experience of many industrial companies that minor accidents involving trifling injuries are more frequent when the lighting conditions are poor. Of course, the poor lighting may contribute directly to the accident, but it seems that as the majority of these so-called minor accidents involve injuries having no exterior visible sign (e.g., bumped head, pulled muscle, back strain etc), there are grounds for thinking that at least a proportion of them are of psychological causation. It is as though the employee is making some sort of cry for help, and certainly cases of clinical depression are of a high frequency where working conditions are poor[(4)]. Unsuitable or insufficient lighting may, of course, significantly contribute to the causation of serious accident, eg those causing death, permanent injury or hospitalization of the injured person (5.1). It should be noted that such serious accidents often have a seriously depressing effect on the morale, and hence the output and profitability, of the whole workforce.

**1.4.7** In this section we have reviewed some of the less easily quantified cost-benefits of good industrial lighting. It must be stressed that improvements in output, accuracy, quality, consistency and other positive factors are the normal responses to removing the handicap of insufficient and unsuitable lighting. This is clear from Table 1, which also indicates that organizations find there is a decrease in accident rate, or a reduction in lost time due to accidents, or a reduction in the minor accident rate which accompanies many re-lighting operations.

Chapter 2

# General, local and localized lighting

For many students of lighting, and indeed for many managers of lighting installations, the practical design steps to providing light at a required illuminance and uniformity in an area are the limit of their interest and ability. Important though these matters are, they are but a small part of the whole subject of lighting. With the present need to exercise stringent control on expenditure and use of energy, these practical matters can be seen as being vital to the success of any scheme. While most industrial interiors are lighted by *general schemes* (2.1) improvements in quality of light may be achieved with *local lighting* and (often with greater economy) by means of *localized lighting* (2.2). Windows are both a blessing and a cause of unwanted ingress and loss of light and heat, and their effects on the energy balance are important (2.3). Finally in this chapter the subject of control switching of lighting is related to those foregoing topics (2.4).

## 2.1 General lighting schemes

**2.1.1** The purpose of a *general lighting scheme* in an industrial interior is primarily to provide for safe movement of the occupants, and to make a contribution to their welfare and comfort, while providing lighting for the tasks to be performed in the interior. This statement needs qualifying by explaining that, where the tasks are not highly visually demanding, and where the tasks are performed over a substantial proportion of the area, then the general lighting installation may satisfy the whole, or substantially the whole, of task lighting requirements in the area. But, if the tasks are visually demanding, then it may be necessary to augment the lighting at work-stations by means of *local lighting* (2.2), and to employ a somewhat lower general illuminance. Further, if the tasks are performed in only a part or parts of the area, and a substantial proportion of the area is used for movement, storage etc, then it may be possible again to light the whole area to a lower general illuminance but to augment the lighting over certain areas where work is performed by means of *localized lighting* (2.2).

**2.1.2** The determination of a suitable illuminance for a general lighting installation usually presents no difficulty, and reference may be made to reliable guidance documents (Appendix II). The usual form of a general lighting installation is for a regularly-spaced overhead array of luminaires at uniform height. But, sometimes, localized luminaires (2.2) may be at a lower height. Special layouts may be needed to provide gradation of illuminance at entrances (2.3), and part of the design task is to determine the most suitable method of switching and control (2.4). If the building is provided with windows in the walls or roof, the interaction of daylight and electric lighting should be studied with the view to determining optimum visual conditions with minimum capital and energy costs (9.1). The general lighting in high buildings or those traversed by gantries may necessitate some variations in choice of luminaire types and their disposition in the space (8.3).

**2.1.3** With the colour-scheme and general decor, the general lighting is a dominant factor of the appearance of a workplace, and some care in the planning to enhance the amenity value of the lighting for the benefit of staff satisfaction is often well justified (18.1).

## 2.2 Local and localized lighting

**2.2.1** The general lighting installation in an interior provides the basic lighting for safety and movement (2.1) but the function of *local lighting* can never be to try and make good the deficiencies of the general lighting system. The functions of local lighting are mainly to boost the general lighting to a value suitable for specific tasks at specific places (2.2.2), and to provide lighting of a special quality or direction for the efficient performance of the visual task (2.2.4). In one important characteristic, local lighting differs from general lighting, in that local lighting is always under the direct control of the operative at the work-station, control of switching, and often control of direction and intensity too. *Localized lighting* is centrally switched, and is not under the direct control of the operative at the work-station (2.2.5).

**2.2.2** The value of local lighting in making significant reductions in energy costs and running costs is simple to demonstrate. Consider the case of a workshop of 20 m × 10 m size, in which visual tasks requiring an illuminance of 750 lux are performed. Lighting the whole area to 750 lux will need approximately 312 000 lamp lumens (assuming a utilization factor of 0.6 and maintenance factor of 0.8) (see 16.1). But, if the visual tasks requiring 750 lux are performed only in certain small areas within the room, a general illuminance of say, 300 lux could suffice, with the illuminance over the workstations boosted to 750 lux by means of local lighting. For example, if four benches were each to be provided with two 85 W fluorescent tubes, the total lumens required in the room would be 125 000 lumens for the general lighting, plus 8 × 6200 lumens = 49 600 lumens for the local lighting, 174 600 lumens in all. This is a saving of around 44 per cent of the lumens, and, in a practical case the energy saving, and the capital cost saving, would be of the same order. This use of local lighting must be applied with care, and the following points must be observed:

(a) The general illuminance should not be less than one-third, and preferably not less than one-half, of the local illuminance at the work-stations.

(b) The colour-appearance of the two systems of lighting should be the same or as close to each other as possible. It is possible to have an overhead general lighting system using HID lamps and to use fluorescent tubes for the local lighting, but the better the match of colour-appearance the better will be the acceptability of the system. If the colour difference is marked, the operator may suffer annoyance and confusion because of presence of multi-coloured shadows.

(c) At the work-stations that are lit by the local lamps, the luminances in each operator's field of view must not be in excessive contrast. The *CIBS/IES Code*[5] recommends that, whatever the horizontal illuminance on the working plane, the ratio of illuminances on walls, floor and ceiling surfaces should be in the following ratios to the task illuminance:

**Table 2** Ratios of illuminance in an interior

| *Location* | *Relative illuminance* | *Relative reflectance* |
|---|---|---|
| Task | 1.0 | Immaterial |
| Ceiling | 0.3 to 0.9 | 0.6 minimum |
| Walls | 0.5 to 0.8 | 0.3 to 0.8 |
| Floor | 1.0 | 0.2 to 0.3 |

**2.2.3** The *CIBS/IES Code*[5] recommends that the reflectance of the immediate background to the task should preferably be in the range 0.3 to 0.5, a level of reflectance that is difficult to achieve in some factories. It also suggests that the night-time average reflectance of walls might be enhanced by use of blinds, but this is entirely impracticable in many factories because of fire-risk, difficulty of keeping such surfaces clean, etc. Multiplying out the relative illuminances by the relative reflectances in Table 2 suggests that, for any task illuminance, the ceiling should have a relative luminance of 0.18 to 0.9, the walls of 0.15 to 0.64, and the floor 0.2 to 0.3, conditions which might be difficult to maintain for long in typical factories (1.2.4). Practical experience indicates that the visual conditions under local lighting should be arranged so that the brightness (luminance) of the task should not be greater than 10-times that of the general surroundings, and preferably with the immediate background of the task having a luminance of 30 to 50 per cent of the task luminance. This ratio of brightnesses of 10:3:1 is about the limit that can be tolerated for long, and a ratio of 10:5:2 will be far more acceptable. A suitable brightness ratio between task and immediate background can be arranged with a backboard or wall behind the task being of suitable colour (and kept clean), and being washed either with the spill-light from local lighting, from the general lighting, or even by special lighting provided to lighten this background to the visual task (3.2.6(a,b)).

**2.2.4** While local lighting can be employed as means of economizing in capital running costs of lighting in certain circumstances (2.2.2), the more usual reason for using local lighting is to give the operator control of the lighting, in both direction and intensity, and possibly in other characteristics too. A local lighting unit having a directable reflector on a hinged, bending or swivelling arm, may be directed horizontally or in other directions needed, enabling the operator to examine cavities, and to get light coming to the task at glancing angles to reveal surface texture—a subject to be examined in greater depth under the heading of *Inspection Lighting* (Chapter 3). If this ability to re-direct the light about the task area is not required, then it may be possible to design an overhead lighting system that does not have adjustable units, but which provides enhanced illuminance over selected parts of the whole room— this *localized* lighting contributing also to the general illuminance (2.2.4). If the lighting of the task will be satisfactory without providing adjustment under the operator's control, then one of the disadvantages of the system may be avoided—namely, an operator will not be able to direct light in a direction which would cause *glare* to other occupants of the space.

**2.2.5** A localized system of light works best if the work-stations are grouped in areas or lines in the room, and the overhead lighting is arranged in a generally symmetrical pattern, but the light over the work-stations is enhanced by one of several means available, such as:

(a) spacing the lines of luminaires closer together over the work-stations,
(b) keeping the lines of luminaires at uniform spacing, but spacing the luminaires in selected lines over work-stations closer together,
(c) lines of luminaires over work-stations may be mounted lower,
(d) making the luminaires over the work-stations of higher lumen output by (i) having more lamps per luminaire, or (ii) higher powered lamps at those locations, or
(e) using a combination of several of these means of getting enhanced illuminance at the work-station areas.

In the case of general lighting which is augmented at work-stations by local lighting, the operatives may switch the local lighting on and off at will, for the contribution the local lighting makes to the general illuminance is negligible, or at least not to be relied upon by others. In the case of localized lighting, the whole installation must be switched as an entity, for extinguishing local lighting would reduce the level of general illuminance needed by other occupants for safe movement and amenity. There is yet one other compromise: to provide a suitable general lighting system for the whole area, and then to add additional groups or lines of luminaires over selected areas. When these additional lamps are in use, there may be a degree of bonus light near the preferentially lighted area, and this will reduce the overall cost-effectiveness marginally; but the system is very suitable when certain operations are only intermittently performed, and thus to light those work-stations continuously to the higher illuminance would be wasteful.

**2.2.6** For departments laid out with lines of benches or machines, it is sometimes possible to localize the rows of lighting over the lines of work-stations, and so arrange things that the spill-light from these luminaires satisfactorily lights the gangways between the work-stations. This system works well where the workbenches or machines occupy a high proportion of the total floor space. Indeed, at high density of floor occupation, there is only a notional difference between such a scheme and a general lighting scheme with an unusually low mounting-height for the luminaires. Installations of this kind are known in laboratories and in the clothing industry, with opaque-reflector trough luminaires being mounted with their base openings below the eye-level of a standing person, possible because all luminaires are mounted over fixed floor obstructions; but very great care has to be exercised in the use of this kind of design to avoid discomforting or disabling glare to seated operators, perhaps by the use of luminaires with base-louvres or other means of restricting the brightness towards the eyes of the seated operatives. With such schemes, the opaque trough reflectors may be apertured to allow a fair measure of upward light (thus preventing the gloomy *'tunnel effect'* resultant from a dark ceiling). Alternatively, provide a small proportion of totally-indirect luminaires (ie inverted luminaires) to throw sufficient light on the ceiling to achieve a pleasant-looking working environment and to reduce the brightness-ratio between the luminaires (as seen by the operatives) and the ceiling and upper walls (2.2.3).

**2.2.7** In certain situations, instead of fitting a local luminaire, it may be safer or more convenient to use a mirror to re-direct light from general or localized lighting to the object which requires local illumination (3.2.6a). Mirrors for this purpose are less vulnerable to damage and less obstructive to the operator than local lighting units, and carry no electric shock risk. Metal mirrors are of course almost indestructible. Use may be made of *'remote local lighting'*, in which a narrow beam-angle spotlight is placed high up in a safe position, directing its light to the task area; these may also be used in conjunction with mirrors near the point of work to direct the light into areas where otherwise it would be difficult to introduce light in sufficient quantity.

## 2.3 Integration of electric lighting and daylight

**2.3.1** The general lighting installation should take into account the availability of daylight entering the building, and the designer should recognize that admixtures of daylight and electric light may cause complications if accurate colour-judgment is required in the area (Chapter 4), and that the provision of windows admitting daylight can be a mixed blessing because of the glare which may be produced (5.2), and because of the uncontrolled heat-gain from the sun's rays (termed 'insolation') (9.2), as well as the heat losses through glass in the winter (9.2).

**2.3.2** Where there are entrances through which workers have to pass frequently between zones of substantially differing illuminances, a decision should be made as to any need to graduate the illuminances at such entrances or at the junction between areas lighted to greater or lesser illuminances. The switching arrangements (2.4) may need to be different by day and by night to produce a zone or zones of intermediate illuminance and thereby reduce the eye-strain, confusion and possible danger which may arise to persons traversing between the two differently illuminated zones (Figure 2.1).

**2.3.3** Although many buildings use a mixture of daylighting and electric lighting (18.1.3), the environmental luminance patterns produced by the two are different, and the difference, as pointed out by Mark Wood-Robinson[(6)], has never been fully explained. The classic example of this is that 'bad light stops play' in cricket at about 1000 lux; yet, indoors, cricketers happily play at 750 lux. Workers in a workplace devoid of natural light but provided with adequate electric lighting express satisfaction with their environment; but other workers, provided with similar electric lighting but having sight of some daylight (possibly that entering by a distant window) may complain of being 'deprived of sufficient light', and envy those working closer to the windows[(37)]. Anomalies abound in situations where daylighting and electric lighting both contribute to the illumination; in exterior lighting practice it has been found necessary to provide a considerable overlap in the switching on of

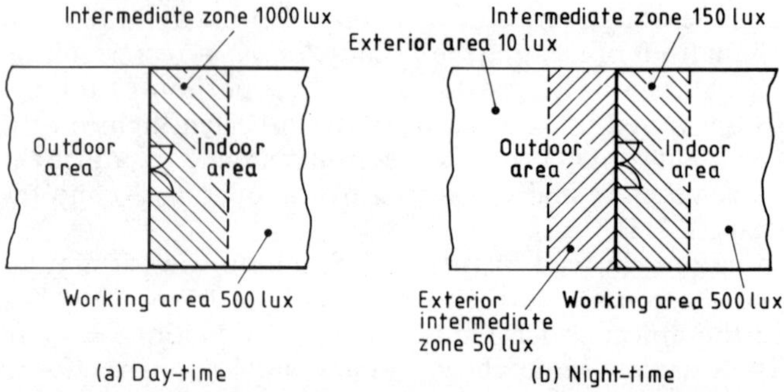


*Figure 2.1* Zones of intermediate illuminance. As an aid to adaptation on the subject's transition between zones of differing illuminances, zones of intermediate illuminance may be provided. (After *Essentials of Good Lighting*[10])

electric lighting as daylight fades[1, 42], while in the lighting of agricultural buildings it has become normal practice to provide a slightly lower standard of general lighting in buildings having windows compared with windowless ones[8].

**2.3.4** Because the illuminance due to daylighting within a building is constantly varying, the amount of daylight admitted is described by the *daylight factor*. The daylight factor is the ratio between the illuminance due to daylight in an interior to the illuminance due to an unobstructed sky (excluding sunlight) at the same instant. Daylight factors in the range of 0.5 per cent up to 10 per cent are specified, and are capable of calculation from data about the room area, the fenestration details and the orientation of the window wall[9]. Typical daylight factors recommended by the *CIBS/IES Code*[5] are: corridors 0.5 per cent; drawing offices 2 per cent; typing and business machine rooms 4 per cent. Where the recommendation is for greater than 4 per cent, it is intended to apply only to task areas, and not necessarily to the whole interior. Bigger windows admit more daylight, ie they produce a bigger daylight factor, and are often viewed as a means of reducing the cost of lighting in a building, though experience shows this not always true (18.1.3). Bigger windows also admit excessive heat and sun glare in summer, and allow escape of costly heat in winter. For working areas with northlight roof windows only, the daylight factor should not be less than 5 per cent, but roof windows which do not face north are not recommended because of their great heat gains in summer.

**2.3.5** Various methods have been devised for combining the use of daylighting and electric lighting, the most well known being *PAL* and *PSALI*. PAL (permanent artificial lighting) more or less ignores the effect of the windows, and the contribution from daylight is treated as a bonus. PSALI (permanent supplementary artificial lighting in an interior) is a method of supplementing the daylight as required according to the conditions from hour to hour and through the seasons, and involves reducing the electric lighting in the areas that are adequately provided with daylight (Figure 2.2). Regulation may be effected by switching out certain luminaires, by switching out selected tubes, or by dimming, by manual control (2.4) or photoelectric control (2.4.6).

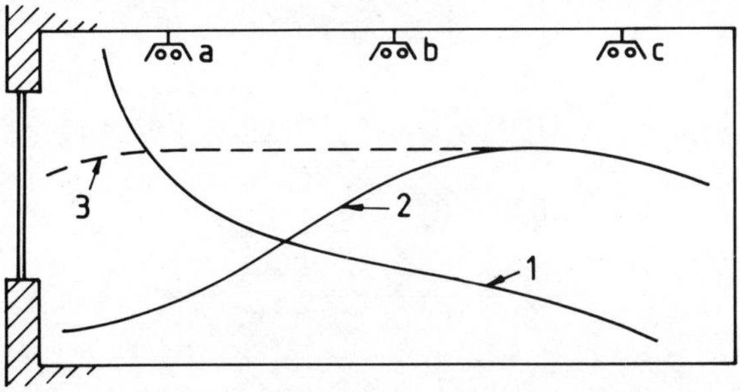


*Figure 2.2* PSALI: daylight is supplemented with luminaires b and c during the day, but luminaires a, b and c are used when daylight is insufficient. Curve 1: typical daylight illuminance at desk level; Curve 2: illuminance due to luminaires b and c; Curve 3: modification of Curve 2 when all luminaires are in use. (After *Essentials of Good Lighting*[10])

## 2.4. Manual and automatic control of lighting

**2.4.1** The arrangements for the manual or automatic switching on and off of a lighting installation form part of a design for a scheme of lighting. Even when no special switching requirement is asked for in the Outline Lighting Specification (7.2), decisions have to be made to switch luminaires either by groups, by lines, by departments or rooms, and conceivably by whole buildings at a time. Not only must the grouping of luminaires to each switch or group of switches be thought about, but the location of the switches also. Without such planning, the process of switching on and off may be time-consuming, the operative having to retrace his steps as he proceeds from switch to switch. In general, operating a switch should not risk placing in darkness persons who cannot be seen from the switching position unless there is pilot lighting or emergency lighting (Chapter 6) sufficient to protect them from dangers in that environment (2.3.4).

**2.4.2** Where it has been decided to spread the lighting load over two or three phases, eg to reduce the risk of stroboscopic effect (5.3), the switching arrangement must be suitable; for example, the luminaires wired from each phase will not be by rows, but will be in a staggered pattern so that each point on the working plane receives light in substantial proportions from luminaires fed from each of the two or three phases (Figure 2.3).

**2.4.3** To prevent waste of energy, lighting in areas that are not occupied should be switched off, or switched down to a level suitable only for safe movement. The question is often asked, *Is it economic to switch off fluorescent tubular lamps and HID lamps for short periods?* to which a general answer can be given as follows: (1) As regards fluoresent tubes of normal sizes used for area lighting, they will probably give normal life if switched not more frequently than once for every three hours of operation throughout their rated life. There is no truth in the idea that the 'inrush current' is so great on switching on that the economy of a short switched-off period will be negated. Fluorescent tubes that are switched on and off several thousand times may blacken prematurely at the ends, and the starter-switch (if used) may need to be replaced rather more frequently (13.4). (2) As regards

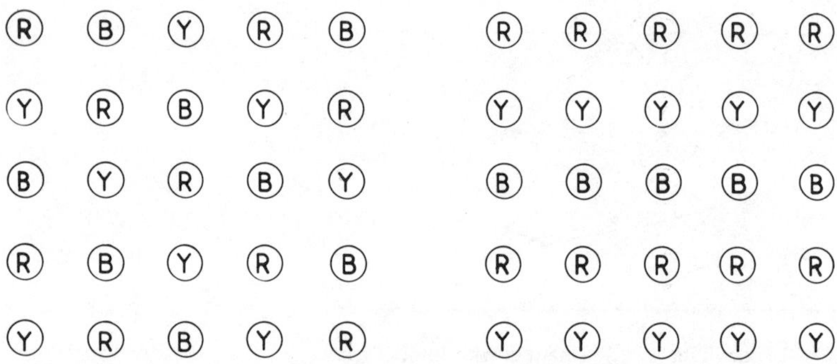


(a) (b)

*Figure 2.3* Spreading the lighting load over three phases of a supply. The method (a) attempts to bring light of all three phases to each point on the working plane. A less costly method (b) attempts to bring light from at least two phases to each point, and is usually satisfactory for all practical purposes.

HID lamps, it must be remembered that, unless wired into special (and rather more costly) control gear, these lamps take some 6 to 10 minutes to cool down before they will re-strike, and this is followed by a further similar period before they run up to full light output again. Thus, very short periods of switch-off are not convenient (2.3.4).

**2.4.4** In factories where energy-saving is attempted by switching off the lamps in unoccupied departments during tea-breaks and lunch-breaks, the economy is small, and care has to be taken to avoid risks to personnel. If, for example, the lights were switched off in a department, and an employee returned there (even unauthorized, or contrary to works rules) and had an accident, it is likely that the employer would be found guilty of a charge of 'failure to provide sufficient and suitable lighting' (Appendix I). In the case of HID lamps, the lamps would have to be switched on again some minutes before the department was to be re-occupied to give time for the lamps to 'run up' to full output.

**2.4.5** Bridging the run-up time for HID lamps may be achieved by providing some ancillary lamps (of a kind that give instant full light output, e.g. tungsten-halogen or tungsten-filament lamps) which will come on with the lamp, and extinguish automatically after a preset period or when the main lighting is at a sufficient level. In one pattern of high-bay luminaire, a tungsten-halogen lamp is incorporated which is switched off automatically by a solid-state circuit when the current taken by the main HID lamp reaches its stable running magnitude.

**2.4.6** Photoelectric controllers are used to control lighting circuits, for example when it is desired to integrate the electric lighting with available daylight in quest of reduced heat-input to the building and savings in cost and energy (2.3). Sometimes time-switches are used to turn lights on and off at predetermined hours, or a combination of time-switching and light-level switching may be employed to give flexibility (e.g. cut-out at weekends) and local response (e.g. during periods of poor daylight, eclipses of the sun). It cannot be too strongly emphasized that any system of remote control or automatic control of lighting circuits is accompanied by an inherent danger that someone could be working on the circuit when it unexpectedly becomes 'live'. Not only should circuits be positively locked off, but, if the remote control cannot be negated, a 'permit-to-work' system should be instituted to safeguard persons completely.

**2.4.7** The control of switching of general lighting (2.1) and localized lighting (2.2) in a workplace should be by a responsible person. Control of the switching of local lighting (2.2) may be left to the operatives that use it; thus, it is important that no reliance is placed on local lighting for its contribution (if any) to the general illumination of the workplace.

**2.4.8** Thought must be given to the effect of switching off general lighting, as the event may trigger the automatic switch-on of the emergency lighting if the circuits are not correctly arranged. However, during the hours of occupancy, when daylight is poor or not available, such an automatic switch-on of emergency lighting could be a requirement for safety on the accidental or unauthorized switching-off of the general lighting. Normally, emergency lighting is switched on upon failure of the mains supply only, but it can be readily interlinked with the normal lighting if this will make for greater safety of the occupants in particularly hazardous situations. Alternatively, a system of 'pilot lighting' may be provided, supplied from the normal supply, and arranged to come on at any time the normal main lighting is off. The pilot lighting will satisfy the requirements within the building for night security, and facilitate the safe movement of the security patrol or fire patrol. On failure of the normal supply, the pilot lights could be switched automatically to the auxillary supply (Chapter 6).

**2.4.9** Switching arrangements can be an important part of energy conservation in the factory. Dr Peter Boyce has made a study of the possibilities[(6)]. On the basis of observations made (actually in an office block) where the luminaires had individual pull-switches in series with group switching at readily accessible and comprehensible switch panels, he concluded that switching patterns are influenced by the motivations of the individual employees. He suggests that it would be beneficial for energy savings if all lights were automatically switched off at lunchtimes, and if a minimum pattern of lit luminaires was automatically established at the commencement of each working spell, because lights once on tend to remain on, whether needed or not. These comments relate particularly to situations where the electric lighting is being used under employee control to augment daylight (2.3).

Chapter 3

# Lighting for difficult visual tasks

Given the advantage of suitable lighting, the human eye can perform far better than is commonly assumed, readily resolving objects of good contrast down to 0.25/0.05 mm without optical aids (3.1). The guidelines in this chapter first direct attention to what may be achieved by the simple use of conventional good lighting, and relate this to difficult and responsible tasks which are routinely performed in industry daily (3.2). The eye's performance can be *assisted* or *extended* to perform tasks beyond its normal capabilities (3.3), and other techniques are employed to enable vision in situations that are physically impossible to see by other means (3.4).

## 3.1 Normal abilities of the eye

**3.1.1** The normal human eye, under conditions of good contrast and adequate illuminance, can resolve small details down to about 1-minute of arc subtended at the eye. Direct measurement of such a small angle presents difficulties, but the angle $S$ subtended at the eye by a small object can be calculated by:

$$S = \frac{3435 \times \text{Size of object (mm)}}{\text{Viewing distance (mm)}}$$

where $S$ is in minutes of arc.

**3.1.2** The *acuity* of the eye is calculated by:

$$\text{Acuity} = \frac{1}{S}$$

and is dependent from instant to instant on the contrast in the object and between the object and its immediate background, the illuminance, and the basic acuity of the subject. (This is the basis of the *Snellen Chart* used for sight-testing by opticians.)

**3.1.3** The required illuminance for a standard of acuity necessary to resolve the fine details in a visual task depends on the contrast (3.1.2). Thus, in visual tasks of poor contrast, more illuminance must be provided to achieve a required standard of visual performance. The required illuminance $E$ to enable the normal eye to resolve a detail of angular size $S$ can be calculated by:

$$E = \frac{193\,700}{R \times S^{1.5}}$$

where $E$ is the illuminance in lux, $R$ is the reflectance factor (between 0.1 and 0.8) of the lightest part of the critical detail in the visual task).

The required illuminance can be arrived at without calculation by use of the nomogram (Figure 3.1), where an example shows that an object subtending 5 minutes of arc to the eye at the normal viewing distance, and having a reflection factor $R$ of 0.2 will require an illuminance of 700 lux for the subject to see it well. A factor of 1.5 should be applied to the resultant illuminance $E$ if there are adverse circumstances, e.g. if the consequences of an error would be costly or dangerous, if the subjects are over 40 years of age, if the time available for seeing is fixed and brief or if the object is moving, if the subject must wear protective goggles etc. Calculations of this kind, modified by extensive field experience, have been used to decide the recommendations given for the illuminance of tasks generally (Appendix II).

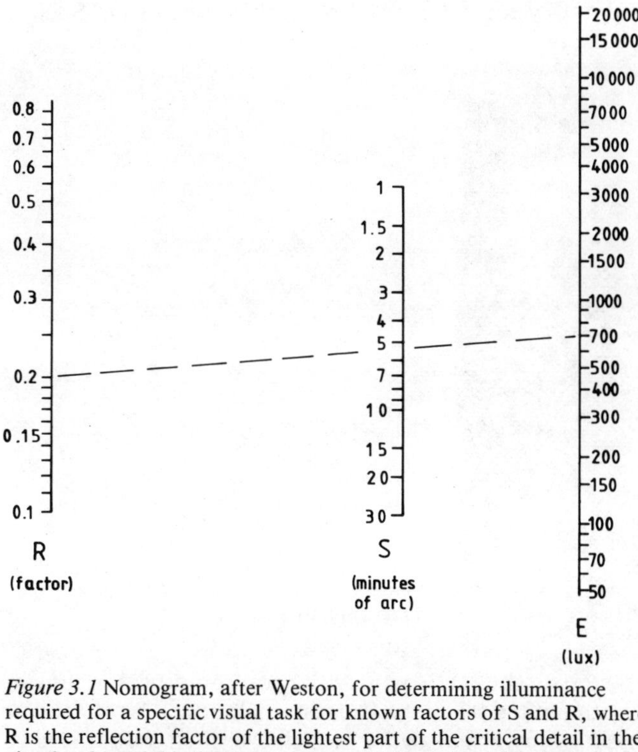


*Figure 3.1* Nomogram, after Weston, for determining illuminance required for a specific visual task for known factors of S and R, where R is the reflection factor of the lightest part of the critical detail in the visual task, and S is the apparent size of critical detail in minutes of arc.

**3.1.4** Consideration of Figure 3.1 will show that for a given factor $R$, the smallest detail that can be seen becomes smaller as the illuminance is increased. This is the so-called 'magnifying effect' of improved lighting, and, where the object is in fact capable of being seen by improvement of illuminance (or of improvement in contrast factor R), this will be generally preferred in the industrial situation to the use of magnifying lenses. In companies where really difficult visual tasks are performed in production, or where there is a constant need for fine and accurate inspection of small detail, the simple application of enhanced general lighting and local lighting (Chapter 2) can do much to improve the speed, ease and quality of visually-based decisions. Only when the possibilities of improvement of vision by these simple and reliable means have been explored (3.2) should more sophisticated methods of providing for the difficult visual tasks be explored, such as *assisted and extended vision* (3.3) and certain *special inspection lighting techniques* (3.4).

An employee whose vision is satisfactory for one task, may not have vision suitable for another task in which the range of contrasts, focusing distances and speeds of objects are different. There is therefore a need for a special type of eye-examination which, if necessary, can be carried out on the work premises by a visiting tester, or by the Company nurse or other trained person. Such a test is termed *vision screening* (3.1.5)

**3.1.5** Vision screening is the process of testing a subject's eyes, using an apparatus such as the Keystone Vision Screener® (Figure 3.2). Such a

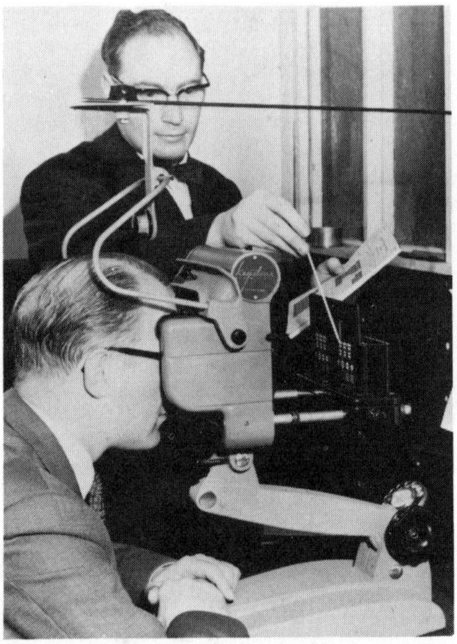

*Figure 3.2* The 'Keystone Vision Screener'® in use. (Photo: Warwick-Evans Optical Co. Ltd, London)

machine enables basic tests to be carried out with great rapidity, and may be used by an operator with limited training. The tests include (1) *Test of usable vision.* Does the subject have normal acuity in each eye when they are tested individually? (2) *Test of stereopsis.* A subject who cannot judge distances is a danger to himself and others in the presence of moving machinery—and the more so if he is required to drive a vehicle or operate a crane. (3) *Test of acuity at required distances.* An operative who can count the legs on an ant at 3 m distance might not be able to focus accurately on a micrometer grauge held at 300 mm from his eyes. Tests will reveal whether the subject can focus at specific distances, e.g. at around 400 mm for VDU operators and typists. (4) *Test for colour vision.* Only specific testing will reveal if the subject has a defect in colour vision. While the possession of good colour vision is an obvious necessity for airline pilots, painters and designers, it may not be realized that a bench-wirer or electrician working with coloured cores can make costly, and potentially dangerous, mistakes by having slightly defective colour vision, even though he may have no difficulty in seeing the red, amber and green of traffic lights. (5) *Other tests.* Tests can be performed for detecting and evaluating the effect of other eye defects, such as 'tunnel vision' (where the subject has defective peripheral vision), 'nystagmus' (where the subject's eyes do not rest steadily on the point of vision, but hunt or vibrate about it), 'night blindness' (where the mechanism of dark-adaptation is defective and the subject cannot see at all in low illuminances), as well as for specific pathological conditions of the eye. Vision screening does not enable corrective lenses or orthoptics to be prescribed, but it does enable a trained lay worker to identify any subjects whose vision is not good enough for the demands of the visual tasks proposed for them, and enable them to be referred for professional advice.

**3.1.6** Very few people have true 'perfect vision', even those who wear properly prescribed spectacles. The variance in their visual performance from the theoretical norm is of little or no importance until they are required to perform visual tasks which stretch their visual abilities to the limit. In most cases, the simple provision of good lighting to the recommended standards of quantity and quality outlined in this book and epitomized by the *CIBS/IES Code*[5] will enable them to perform to a satisfactory standard and without suffering eye-strain. However, increasingly in the context of modern industry, subjects are required to perform tasks which involve a degree of eye co-ordination which have not been commonly needed previously. This is one reason why vision screening tests (3.1.5) are so important, particularly in checking binocular vision and eye co-ordination in addition to the normal tests for stereopsis. This faculty of vision is important for persons operating VDUs and similar equipment, and for those who have to work with or observe fast-moving machinery, where any deficiency in binocular vision can cause considerable eye-strain; in both instances the visual deficiency may reduce operator efficiency and detract from job satisfaction, while in the second case it could also be a factor of accident causation (5.1).

**3.1.7** There are a few production tasks in which there is a requirement for the continuous performance of very difficult visual tasks, e.g. diamond brilliant cutting, matching of printing inks, work on micro-electronics etc; but, in general the most difficult visual tasks are concerned with the processes of inspection. All the information relating to inspection in the remainder of this chapter will apply equally to production tasks of high visual difficulty.

**3.1.8** The classic work on the subject of lighting for inspection is a paper by Bellchambers and Phillipson[11] published in 1962. Of course, technology has moved on considerably since then, but the principles do not change. The basic approaches of these authors hold good, including their convenient classifications of *direct vision* (also termed unassisted vision) (3.2), *assisted vision* (3.3) and *extended vision* (3.4) (Figure 3.3).

## 3.2 Inspection by direct vision

**3.2.1** Amongst production engineers, the term 'inspection' is strongly associated with mensuration and the use of test instruments; but a review of

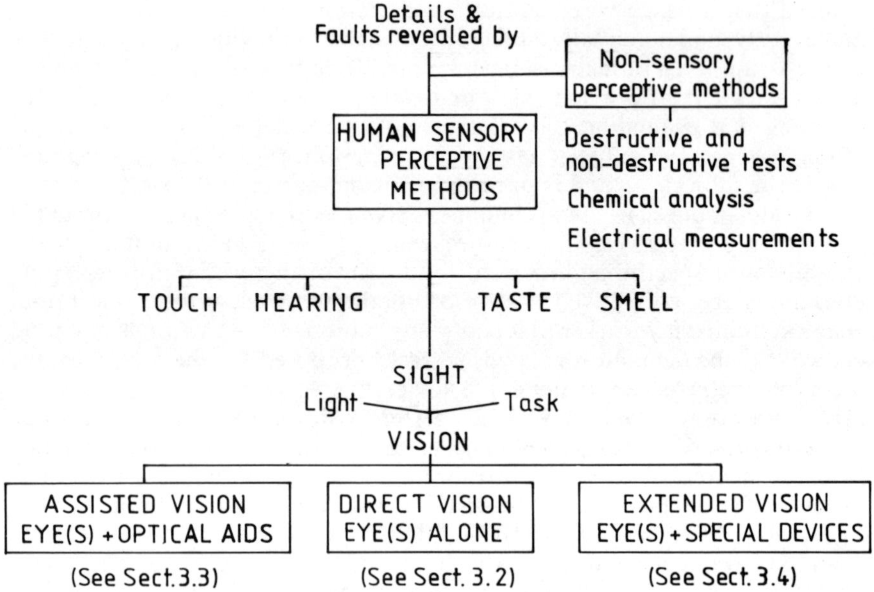


*Figure 3.3* An analysis of the methods of revealing detail and faults in an inspected object or product. (After Bellchambers & Phillipson[11])

common inspection tasks in many manufacturing industries shows that most 'inspection' is performed by the unaided use of the inspector's eyes, i.e. it is inspection by direct vision. (The term 'direct vision means that no optical aids are used other than the wearing of refractive lenses prescribed by the optician). It is obviously important that the inspector's vision shall be of a satisfactory standard. To this end, it is common these days for staff to be asked to submit to an eye-examination by a qualified optician, a course to be recommended particularly in the engagement routine of new employees. The visual performance of a subject is a complex of abilities, but an adequate assessment may be made by vision screening (3.1.5).

**3.2.2** Given a person having a satisfactory standard of vision, training and practice may enable that person to achieve remarkable performance of accuracy, consistency and speed in visual inspection tasks. Although an inspector may appear to be 'just looking', in fact he or she may be exercising a very high degree of skill, and contributing significantly to the quality-control of the organization. With practice and training, a skilled inspector can exercise such a high standard of vigilance, alertness and keen observation, that the probability of a significant fault being overlooked is very low. This high standard of inspection quality may be accompanied by very rapid performance, so that the cost of inspection may be as low as a few per cent of the manufacturing labour cost. Inspectors cannot be expected to achieve and maintain high standards of accuracy and speed if the visual conditions and the working environmental conditions are adverse. Of all external factors, lighting has the greatest effect on the performance of inspectors. Before analysing how the lighting and environmental conditions may be tailored to aid inspection (3.2.6), the economic importance of inspection will be examined (3.2.3, 3.2.4, 3.2.5).

**3.2.3** Reputations of companies are founded on the quality of their products; the satisfaction of their customers, and indeed the very safety of their customers, depends on the goods being 'of merchantable quality and fit for the intended purpose' as required by the UK *Sale of Goods Act*. The Act embodies the fundamental principle that if the seller is to avoid culpability should the user suffer loss, damage or injury, he must show that the goods were of merchantable quality etc. It will also be evidence in his favour if he can show that he took steps to protect the buyer from faulty goods, for example by instituting effective quality control, testing and inspection, and kept adequate records of these steps. According to the type of goods and their application, goods will be subjected to 100 per cent inspection, or a statistically representative sample will be inspected before despatch to the customer. Although the losses might be mitigated by costly insurance, the principle of 'manufacturer's liability' can bring ruin on a company which is the subject of claims; laws in USA, and in UK and other EEC countries are gradually becoming harmonized on this matter. Thus, the ultimate viability of a company might depend on its inspection department. As has been stated, most inspection will be carried out without use of instruments, magnifiers or other apparatus other than good lighting (3.2.5). *Assisted vision* (3.3) will be needed for some visual tasks, and for others the techniques of *extended vision* (3.4) will have to be employed.

**3.2.4** The philosophy of the 'division of labour' is carried so far in some factories that is not unknown for a production worker to fail entirely to inspect his own output, saying that 'inspection is the work of the inspection department'. In carrying out studies of inspection problems and their lighting needs in factories (as well as helping with the associated management problems) the author has often found that lighting in production areas was insufficient for production-line workers to inspect their own output adequately. In some cases, simply raising the general lighting level to the *CIBS/IES Code*[5] (and providing such local lighting as might be necessary) has enabled production-line workers to observe departures from dimensional or quality requirements quickly, and thus cease production of further defective goods—thereby significantly reducing the scrap-rate. In some cases it has proved economic to disband a separate inspection department, and to distribute the inspectors in the production area, some as 'roving inspectors', and some at 'inter-process inspection stations'. (This cannot be done if special environmental conditions are required for inspection, e.g. controlled ambient temperature or humidity, or if 'clean-room conditions' (10.3) are required; neither can it be used if the inspection processes use costly equipment that would have to be duplicated at several inter-process inspection stations). Dispersed and roving inspection is very successfully used in the clothing industry, in the furniture industry and in the manufacture of boots and shoes, and is applicable to precision engineering and fabricating trades. This routine of inspection can be economic and good for the organization, but the management aspects must be carefully handled to reduce potential antagonism between inspection staff and production staff, and to bring about a realization that inspection is not a separate process, but is part of the process of production. There is another potential bonus, in that it is possible to reduce greatly the cost of transporting goods between the stores, work-stations and the central inspection department, with all the attendant clerical costs (Figure 3.4). The first requirement is that the lighting in the production area must be sufficient and suitable for inspection purposes.

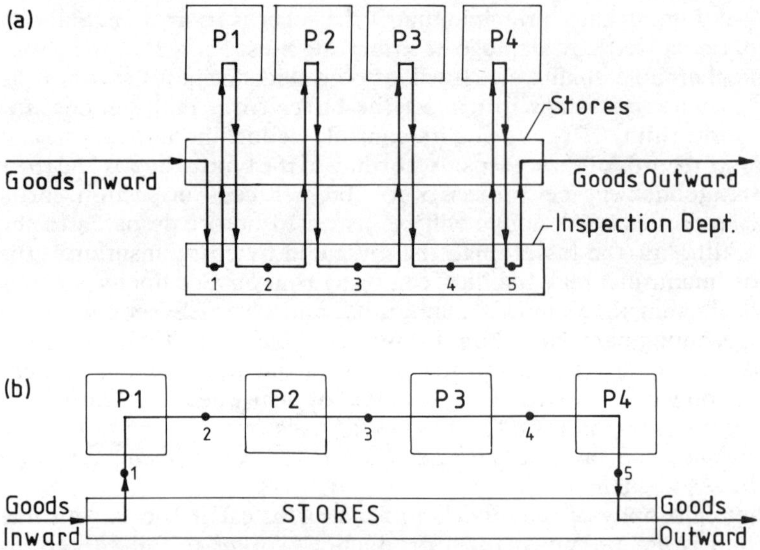


*Figure 3.4* Central versus dispersed inspection in a factory (simplified). (a) Conventional flow, with central stores and central inspection department. The four manufacturing processes (P1, P2 etc) and five inspection processes (1, 2 etc) require 20 stores transactions (shown by arrow-heads) and considerable time and cost for moving the goods to and from the central stores. (b) Linear flow, with dispersed inspection stations. The four manufacturing processes (P1, P2 etc) and five inspection processes (1, 2 etc) require only 4 stores transactions (shown by arrow-heads) and the time and cost of moving goods has been greatly reduced. This system is only possible if the lighting conditions in the production areas are good enough to permit the performance of the inspection processes. The space needed by a conventional inspection department is also saved.

**3.2.5** As an illustration of the effect of improvement of lighting on rates of fault-creation and fault-discovery, consider the case of a company producing domestic applicances which included some zinc-alloy die-castings. At the beginning of the consultancy (which had been instituted because of the high frequency of customer complaints about die-cast components which fractured in use) the general lighting in the production and inspection areas was measured and found to be 180 lux and 270 lux respectively. It was considered that these illuminances were insufficient to reveal the fine hair-cracks, crizzles and porosity in the die-castings. The fault rate (as determined by the inspection department) was put at 7.5 per cent, but it was apparent from the level of customer complaints that a significant proportion of faulty castings was getting through the inspection process. As a first step, the inspection department was relighted to 500 lux, using fluorescent tubes of good colour rendering, and local lighting capable of producing 1000 lux on the inspection task made available at each inspection station. It was quickly established that the fault rate was far higher than had been previously thought; it was determined to be about 17.8 per cent, a figure far too high for economic operation. The company then introduced better lighting in the production areas, putting in a suitable general lighting scheme to produce 500 lux. Within a few weeks, the true fault rate in production fell from 17.8 per cent to around 3.8 per cent due largely to the ability now of the operators to see faults in the goods they were producing, and to take corrective actions. As a result, the reject rate in inspection fell to around 3.2 per cent, indicating that at least 84 per cent of all faults were being discovered during inspection. This still meant that 0.6 per cent of goods sold were faulty, but the management decided to live with this on the grounds that (a) this was such a good improvement over the previous 17.8 per cent fault rate, and (b) experience soon showed that the undiscovered 0.6 per cent of faults were mainly minor blemishes which did not presage failure of the casting in normal use. (In fact, experience through this exercise enabled the management to revise their criteria for 'pass' and 'not pass' defects, on the basis that more accurate assessment of faults was possible with the improved lighting, leading to further minor savings). The nett cost of the improvements to the lighting (which involved replacing outworn fluorescent tubular lamp luminaires with new high-pressure sodium lamp luminaires) was estimated to have been fully recovered by (a) reduction in faulty items, (b) improvement in output, (c) reduction in labour turnover and lost time, and (d) energy savings, in a period of 37 weeks, and thereafter there was a continued enhancement of profitability.

**3.2.6** Given that the operatives are provided with sufficient illuminance and a suitable quality of lighting for their tasks, there are a number of practical steps which may be taken which will further improve the quality of the work, or which may contribute to productivity or safety. Such steps include:

(a) *Enhance the contrast between the object and its background.* Example: in a factory assembling small relays, the girls cut their fingers frequently in picking the components out of fibre tote-boxes. Action taken was to replace the tote-boxes with shallow trays into which the girls could see easily; this not only enabled them to avoid cutting their fingers or running sharp objects under their nails on picking them up, but enabled them to pick them up in the correct orientation for immediate assembly—thereby gaining in work-rate. The insides of the new trays were painted white to aid contrast between the components (which were matt black) and the inside of the trays.

(b) *See an object in silhouette.* Example: in the same factory as that quoted at (a), the relays had to be finally adjusted by hanging them up at the work-station and adjusting the springs and contacts while a weight simulated the pull of the coil. It was difficult to see into the contacts. Action taken was to provide a well-illuminated white board at the back of each work-station, so the mechanism could be seen in silhouette. This simplified the task, improving both output and accuracy.

(c) *Improve the colour-scheme in the workplace.* Example: in a drawing office, neutral light tones on all visible surfaces produced a 'bland field' in which the eye found no resting place. The measured illuminance at drawing-board level was 1200 lux, and the calculated Glare Index was 18.8, fully satisfying the requirements for illuminance and glare limitation, yet there were continuing complaints about the lighting. Action taken was to introduce some small areas of vivid colour into the decor by application of paint to convenient surfaces and the addition of some strongly coloured paintings on selected walls. These paintings were spotlighted with a few low-power projectors on track, and, purely for psychological reasons, the tubes in the overhead lighting were changed to others of a slightly different colour-appearance (but having the same light output and colour rendering). The staff expressed complete satisfaction with the improvements.

(d) *Introduce directional lighting.* Example: in a department where hides and large reptile skins were graded and sorted for cutting, on moving the department to an area with rooflights and no wall windows, there were complaints that the lighting was inadequate, even when measured at over 1500 lux with the mixture of daylight and electric light. Action taken was to provide blinds which could be drawn over selected roof windows on very bright days, and some asymmetric wall-mounted fluorescent tube luminaires were placed at the end of the worktable. This introduced a near-horizontal flow of light, which helped to reveal surface texture and thus aided the work. (In this case, the management rejected a proposal to place the department in another location without rooflights, but it has since been noticed that output and quality in the winter (when daylight is limited) actually *improves*; this effect is no doubt due to the improved $E_v/E_h$ ratio when the light from the rooflights is reduced). Similar effects have been noted in the 'salles' (inspection rooms) in the paper industry, where light at a flat angle to paper sheets aids their inspection for texture (19.2).

**3.2.7** In production and inspection tasks involving critical vision, any means that can emphasize the critical matter in the 'object of special regard' will speed up the process and add to accuracy. Every industry has its special visual difficulties, and many kinds of optical, mechanical and electrical devices are used to reveal particular features of the product. Inspection may be aided by the use of one of the other senses, or non-visual data may be transformed in some way to make it visual or otherwise assimilable by the operator. The lighting may need to be modified, perhaps not only in increasing the illuminance on the task but also by change or control of one or more of its other attributes, e.g. parallel, divergent or convergent beams of light; light from nearly point-sources or from large area diffuse sources; the angle of incidence of the light may need to be controlled (3.2.6); high colour-quality light may be needed for colour discrimination (Chapter 4), or monochromatic or non-full-spectrum light used to reveal one particular colour which otherwise might be difficult to see (19.3.3). These measures of control may be difficult to apply to a whole department and it becomes a practical necessity to create a miniature environment in which the desired lighting conditions can be produced — i.e. to use an *inspection booth* of some kind. There are difficulties relating to adaptation on entering and leaving booths, and particular problems with colour-adaptation (4.4). Some subjects feel claustrophobic on entering small spaces, and there may be difficulties of providing adequate ventilation. Because of these difficulties, these days it is usually preferred to use 'head-and-shoulders' booths, of which several practical patterns have evolved. The most generally used types are those in which the booth is a moveable item, which can be placed on any bench of convenient height. Where it is decided to introduce *dispersed inspection* (3.2.4), such booths can be readily placed within the manufacturing area for inter-process inspection of products. It may be necessary to provide some screening from the overhead general lighting or other lighting, in order to standardize the conditions within the booth, and to enable the inspector to keep to a suitable level of adaptation both to colour and to illuminance (Figure 3.5).

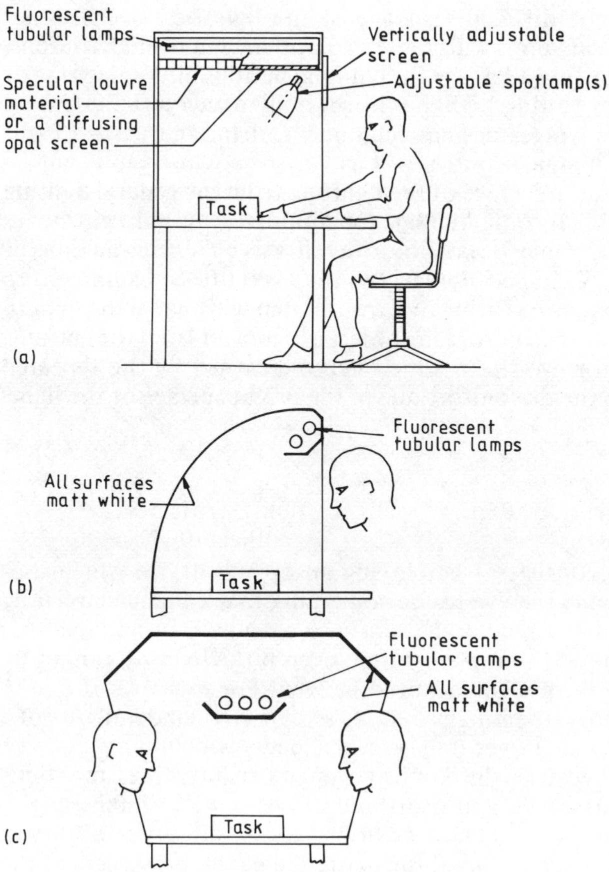


*Figure 3.5* Some types of 'head and shoulders' inspection booths. (a) Using direct lighting from a specular louvre material, or from adjustable spotlamps, or diffused direct lighting from an opal diffusing screen below fluorescent tubular lamps; (b) Indirect lighting; (c) Indirect lighting arranged so that the workstation can be accessible from both sides. (Illustrations (a) and (b) are based upon Bellchambers & Phillipson[11])

**3.2.8** A valuable inspection method is to place a sample of translucent or transparent material over a light-box or similar device. It is a minor doubt whether, strictly speaking, this constitutes assisted or unassisted vision, but the method is extensively used for such diverse operations as checking the positions of stiffeners in shirt collars; inspecting sheets of dried wood pulp for inclusions (such as dead insects) before the material is re-pulped to make into medical filters; examining artists' and photographic transparencies (4.4.11); and the routine in-line inspection of materials such as bottles, pots of jam, ampoules of drugs, watermark inspection in banknotes and securities etc. Light-boxes, containing fluorescent tubes or tubular architectural tungsten-filament lamps have long been used in drawing-offices for tracing drawings on to linen or plastic, and they are widely used for lithographic platemaking from colour transparencies. Care has to be taken to prevent the lamps or the diffusing surface of the light-box overheating; overheating of fluorescent lamps will create a colour-shift of the light output, while general overheating may damage the equipment or its internal wiring or control-gear. The art of getting highly uniform luminance on the diffusing surface of a light-box involves design techniques outside the scope of this book. The fitting of a dimmer-control will usually be advantageous, and it may be helpful to shield the surface of the lightbox from the general lighting so that luminance coming through the sample dominates. Bright light-boxes are a source of glare, and it may be possible to avoid this by placing an opaque cut-out (with its aperture shaped similarly to the object to be examined) to reduce stray light. Further, the fitting of a foot-switch will enable the light to be switched-off or diminished before the object is removed from the cut-out, thus preventing operator eye-strain which may be caused by the repeated exposure of the eyes to the discomfort due to the bright surface of the light-box.

## 3.3. Assisted vision

**3.3.1** Assisted vision consists of employing an optical device which uses visible radiation to enable the eye to see something that cannot be seen by unassisted vision. It must be made clear that such a device is not a substitute for providing sufficient and suitable lighting, even if the lighting cannot by itself enable the eye to resolve the detail to be seen. For example, if due to poor light we cannot generate enough acuity to see a particular detail, use of a magnifier will produce a larger but greyer image which may also be unreadable. This is because the optical system will gather the light over a narrow collection angle and distribute it over a wider image angle, and necessarily that image must be less bright. Also, all lenses absorb a proportion of the light entering them, further reducing the brightness of the image. While it is desirable to increase the illuminance on the task which is to be viewed through an optical system, such increase must be accompanied by an increase in the luminance of the background to the task, or unbearable glare may result on using the optical system. Some examples of assisted vision are given in this section, somewhat extending the information given by Bellchambers and Phillipson[11] and covering *magnifying lenses* (3.3.2), *microscopes* (3.3.3), *telescopes* (3.3.4), *mirrors* (3.3.5), *profile projectors* (3.3.6), *transmitted-light devices* (3.3.7), *periscopes* (3.3.8) and *illuminated grids* (3.3.9).

**3.3.2** *Magnifying lenses.* These are used in many forms, including:

3.3.2a *Watchmaker's glass* (also known as an 'optic'), which is worn by the operator as a monocle, and, of course, gives only monocular magnification. In another form, the monocular magnifier may be mounted in a spectacle-frame, and the optic may be hinged to permit to be swung out of the line of vision. Long periods of work with a monocular magnifier can be tiring, particularly if the same eye is consistently used, because of the eyes becoming differentially adapted. Continuous monocular work can lead to a pathological condition called 'lazy eye', in which the under-used eye becomes defective in alignment and focus.

3.3.2b *Binocular magnifiers* consist of two watchmaker's glasses (3.3.2a) mounted in a spectacle-frame. These are less likely to cause eye-strain than monocular magnifiers, but the optics in front of the operator's eyes effectively 'blind' him for distance viewing, and for safety should only be worn when seated.

3.3.2c *Hand-held magnifiers.* These tend to become scratched in use, plastic lenses being worse in this respect but lighter in weight. Glass lenses will more readily shatter. Because the hand is not steady, the practical limitation of magnification is about ×5, and there will be a fair amount of chromatic aberration from ×3 upward.

3.3.2d *Magnifiers on stands.* These may have a local lamp or lamps built into the bezel and screened from the operator's view. The preferred type of lamp is the miniature fluorescent tube, but where the task involves critical colour discrimination as well as need to resolve small detail, it may be difficult to obtain tubes of small powers in the better-colour-rendering grades of phosphor. The magnifiers may be on flexible or adjustable mountings for ease of adjustment. One pattern includes a small fan for blowing away fumes and smoke from small soldering jobs for which the magnifier with its light is needed.

3.3.2e *Prismatic magnifiers.* Where magnification is required in one direction only (as for reading the delicately thin column of mercury in a clinical thermometer), a prismatic magnifier consisting of a half-cylinder of glass or clear plastic is sometimes used.

**3.3.3** *Microscopes.* Microscopes include all magnifiers having compound lenses with a fairly narrow collection angle, and magnifications from ×5 upward. A number of forms exist, including:

3.3.3a *Bench microscope.* This may be monocular, but these days it is more common for the instrument to be fitted with a prismatic image splitter for binocular vision. Special instruments may provide binocular viewing plus a position for a camera to record the enlarged view, or even for double-binocular viewing (for two operators simultaneously). Coloured light (by use of filters) may be employed, and the object may be illuminated (back illumination or front illumination) by a mirror which reflects light from the local or general lighting (thus avoiding problems due to colour difference which can occur if a microscope-light is used).

3.3.3b *Machine mounted microscopes* for reading vernier scales or for examining the workpiece are built into some precision machine tools.

**3.3.4** *Telescopes.* These are occasionally used in industrial plant so that an operator can see some distant gauge or mechanism without leaving his operating position. If the information cannot be conveniently and economically displayed adjacent to the operating position, a simple fixed telescope will be cheaper and more reliable than the now commonplace method of using a closed-circuit television (CCTV) loop.

**3.3.5** *Mirrors.* Simple plane mirrors have a variety of uses in lighting and inspection work, and generally may be used (a) to direct light, or (b) to reflect an image.

3.3.5a Mirrors are used to direct light in areas where it is inconvenient to place a luminaire, e.g. where the physical space is too confined, or where there is a hostile environment (heat, danger of explosion etc) which prevents the introduction of electrical apparatus; there are thus advantages in using a mirror instead of a local luminaire (2.2). If there is a high probability that a mirror might be scratched or broken in the particular situation, it might be robustly constructed of polished and lacquered metal, or consist of a plate of metal which has been chromium plated and polished. Metal mirrors for this purpose are sometimes lightly matted to give a more diffuse reflection and prevent images of lamps being seen. Combinations of two or more mirrors may be used to direct a greater quantity of light in the chosen direction. The light loss at a mirror is between 8 and 14 per cent (front-silvered mirrors—which are rather fragile—can give losses as low as 4 per cent); they may conveniently be used in combination to direct light through several changes of angle to the desired place. When light has been reflected twice in the same plane, it becomes substantially polarized (App. V).

3.3.5b Mirrors are used to reflect an image to an observer, the mirror being placed where it would be inconvenient or dangerous for the operator to place his eyes—e.g. where the physical space is confined, or where there is heat or danger from moving machinery. Combinations of mirrors are sometimes used to view inaccessible objects (periscopy), and there are methods of using a single system of mirrors to both transmit light to the object and bring back a virtual image to the subject (duplex periscopy) (3.3.8). There are practical limitations on the length of light path and number of reflections that can be effectively used because of light losses in the mirrors (3.2.6a) and due to distortion at the mirrors. Mirrors may be used on inspection lines to see the tops or backs of objects on a conveyer belt, or to get a good view of a continuous web of paper when looking for 'thins' and inclusions (19.2.3) and for colour checking too (4.4.2).

3.3.5c Another valuable application of mirrors in inspection is to take advantage of the fact that when a mirror rotates through an angle $\theta$, the reflected ray rotates through an angle of $2\theta$ (Figure 3.6). Thus, components may be tested for angular accuracy by fixing a mirror thereto, and observing the displacement of an incident ray, e.g. a collimated narrow beam of projected light, or the movement of a virtual image.

**3.3.6** *Profile projectors,* also sometimes known as 'shadowgraphs',

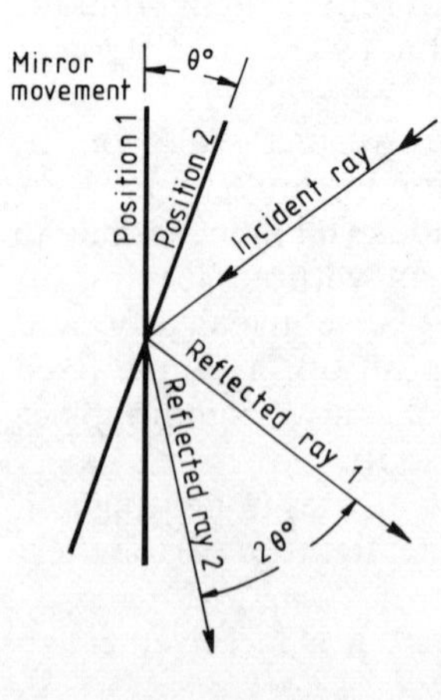


*Figure 3.6* Angular displacement of a reflected ray. When the mirror rotates through the angle θ the reflected ray rotates through 2θ; an effect that can be used to detect displacements of small angular dimension.

employ the fact that shadows thrown by objects in the light of a small source are larger than the object (Figure 3.7). The shadow may be cast on a white opaque screen, or may be viewed from the reverse side of a translucent screen, e.g. of ground-glass, flashed-opal glass, or diffusing plastic material. The magnification effect due to the divergent rays may be enhanced in one direction by tilting the screen to get an elongated shadow. If the positions of the light source, the object and the screen are controlled, the shadow profile may be to a known magnification, and its outline can be compared with a standard outline for correctness. In a more sophisticated version of this device, the shadows are cast by a source having a mirror and condenser glass to produce a near-parallel beam; this beam after being partly occluded by the object is then magnified by a further system of lenses and mirrors to present a silhouette on a translucent screen. The silhouette of a female configuration can be viewed by a profile projector by first making a cast of the cavity in a quick-setting material which does not have much variance in volume between its liquid and solid state, e.g. pure sulphur.

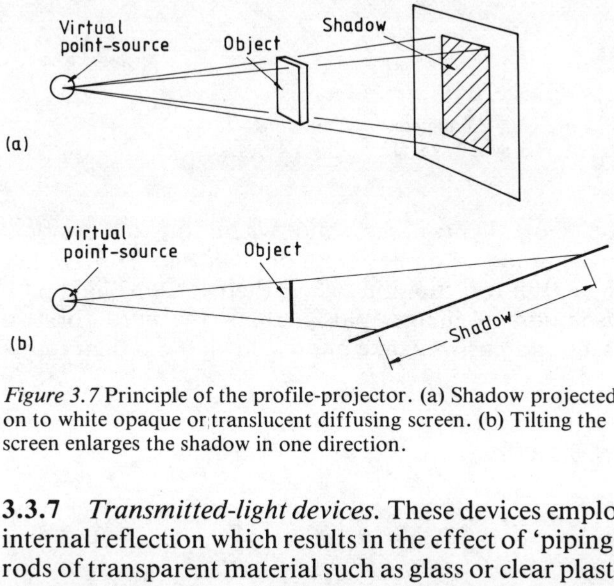


*Figure 3.7* Principle of the profile-projector. (a) Shadow projected on to white opaque or translucent diffusing screen. (b) Tilting the screen enlarges the shadow in one direction.

**3.3.7** *Transmitted-light devices.* These devices employ the principle of total internal reflection which results in the effect of 'piping of light' through solid rods of transparent material such as glass or clear plastic. Such devices can be used to bring an intense light to a small object without creating any scattering of light, and there is no heat transmitted with the light (Figure 3.8). This is a useful technique to produce intense illuminance on a small object without damaging with heat, e.g. in the examination of biological samples. If the rods, say of glass, are reduced to the diameter of hairs (0.25 mm to 0.05 mm diameter) the 'piping effect' still takes place. Bundles of such glass fibres may be used to conduct the light in place of a solid rod, with the advantage of the bundle being flexible. Such a 'fibre optic' may be used to direct light to the bottom of deep narrow apertures, round corners, or into enclosed spaces with small entries. If the fibres in the bundle are arranged so that each fibre at its distal end is in the same positional relationship to the other fibres in the bundle at its proximal end, and the bundle is said to be 'coherent', and is capable of transmitting an image along its length. The resolution of the image will depend on the density (number of fibres per unit area). Fibre optics of sophisticated design can perform in duplex mode, i.e. they can transmit light to their distal end, and an image can be brought back through the same fibres, so they behave similarly to a duplex periscope (3.3.8). A duplex fibre optic could thus be used to inspect the interior of vessels containing flammable substances; for applications without a fire risk, an alternative method is to mount a small lamp at the distal end of the device, and use the fibres only for bringing back the image.

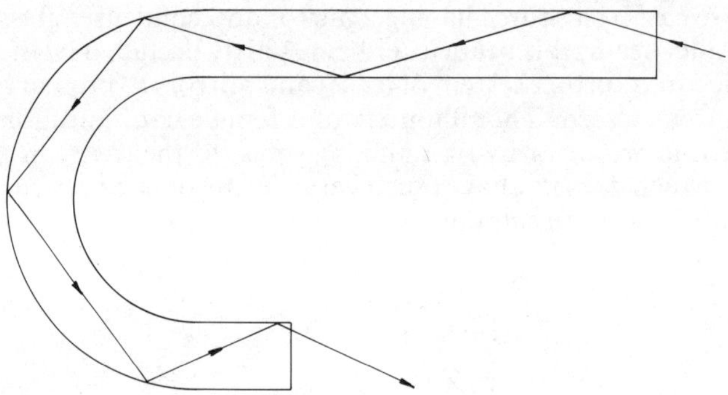

*Figure 3.8* Principle of 'piping of light'. An incident ray, entering the end of a polished transparent rod, will travel along it by a series of internal reflections.

**3.3.8** *Periscopes.* For short light paths, and where the dimensional constraints are severe, the mono or duplex fibre optic devices (3.3.7) have much to recommend them. But for situations where there is a long light path and need for high resolution of image, plus perhaps the need for two observers to see the image and possibly take photos down the system too, a

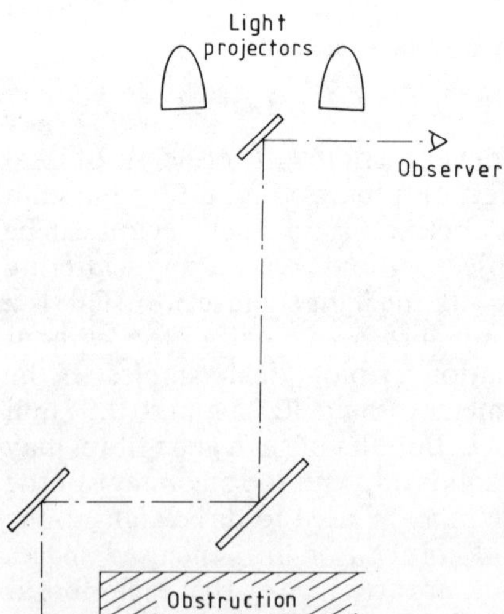


*Figure 3.9* Principle of one configuration of a self-illuminating (duplex) periscope.

plane-mirror periscope will have the advantages of producing a bright high resolution image. Over extended distances the periscope may be combined with a telescope, and it may have an image splitter at its head for dual viewing or photography (Figure 3.9).

**3.3.9** *Illuminated grids.* Grids marked on background boards, or painted on back-lit diffusing panels have several valuable uses in inspection, both for inspecting for flatness (3.3.9a) and for examining transparent materials (3.3.9.b).

3.3.9a *Flatness inspection of specular sheets.* An illuminated grid (typically comprising black lines 2 mm wide in a grid pattern of lines at 15 mm centres) is used to inspect a specular surfaced sheet by observing the reflection of the grid on the surface of the sheet. Because of the 'doubling of the angle' effect (Figure 3.6), small deviations in flatness will be revealed by quite marked deviations in the image of the grid (Figure 3.10). See also Figure 19.4.

3.3.9b *Inspection of transparent materials.* An illuminated grid as described (3.3.9a) can be used as an aid to inspecting transparent objects such as glass bottles or sheets of glass or clear plastic. Because of the varying refractive effects due to variations in the thickness of the material or the lack of parallelism between two nominally parallel surfaces, the grid appears distorted when viewed through the transparent material and faults are more easily noticed (Figure 3.10).

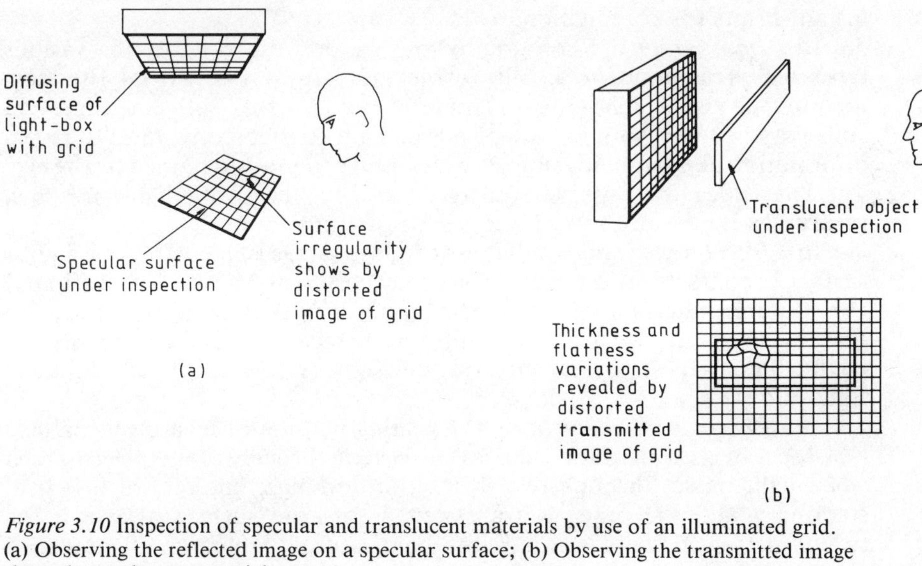


*Figure 3.10* Inspection of specular and translucent materials by use of an illuminated grid. (a) Observing the reflected image on a specular surface; (b) Observing the transmitted image through translucent material.

## 3.4 Inspection by extended vision

**3.4.1** The term 'extended vision' is applied to inspection by means other than the ordinary use of the eyes alone (direct vision) (3.2), or with some form of optical device (assisted vision) (3.3) plus white light. Although these techniques lie outside the subject of this book, some examples of extended vision methods are given here for reference, somewhat extending the range of techniques listed by Bellchambers and Phillipson[11].

3.4.1a *Ultraviolet irradiation.* This is used to create fluorescence in substances as a means of analysis. For detailed physical analysis an ultraviolet spectrometer may be used. UV is also used for crack-detection, where a metal object is dusted with a fine powder of a fluorescent material, or is painted with a solution of such a material. When the surface has been cleaned, irradiation with UV will reveal minute traces of the fluorescent material in cracks and crizzles on the surface of the object, faults which may be impossible to see with the naked eye in normal light.

3.4.1b *X-ray irradiation.* Objects may be irradiated with X-rays and a visible-light image observed which is formed by the rays which having passed through the object impinge on a fluoroscopic screen. By this means internal faults may be detected. Detailed physical analysis of materials may be made by means of an X-ray spectrometer.

3.4.1c *Low light level irradiation.* For substances which would be adversely affected by irradiation with high intensity visible light, examination may be made under controlled illuminance of as low as 0.001 lux, the object then being scanned with a scintillator or photo-multiplier device.

3.4.1d *Infra-red irradiation.* Objects which would be adversely affected by other forms of radiation may be examined by irradiation with infra-red light. An infra-red-sensitive film is used to photograph the object and produce instant-prints for examination.

3.4.1d *Photography.* Photography may be performed under the various types of irradiation, e.g. ultra-violet (3.4.1a), X-rays (3.4.1b), low illuminance visible light (3.4.1c) and infra-red (3.4.1d). Both cine films and stills may be taken with the use of pulsed light (stroboscopy) (5.3) to reveal information about moving objects which move too fast to be seen by the eye. The latest idea is to use a video recorder with the strobe light, for such recordings can be instantly played back at low speed.

3.4.1e *Interference bands.* Flatness of glass surfaces in contact, or a glass surface in contact with a metal surface may be assessed by the observation of the coloured interference bands (Newton's Rings) created by diffraction occurring between two nearly optically flat surfaces. Gratings can be used for measuring flatness by a method devised by the National Physical Laboratory.

3.4.1f *Photoelectric detection.* The grating method of measuring flatness (3.4.1e) uses as a detector changes in current through a photoelectric cell when light passes through two diffraction gratings, one of which is held stationary and the other is translated across the surface of which the straightness or flatness is required to be measured. Similarly, alignment of gaps, holes etc can be checked by projecting a light beam and detecting it with a photocell.

3.4.1g *Visible light of colour other than white* may be used for certain difficult visual tasks. The well known example is that used in printworks to align accurately the register marks for yellow, the yellow marks being more clearly visible under blue light obtained usually by a cyan filter held before the eye or placed over a tungsten-filament lamp local luminaire (with the general lighting subdued) (19.3.3).

3.4.1h *Polarized light.* Studies of stress patterns in objects can be made by making an exact copy (full size or to scale) of the object in a clear plastic material. This is then observed by transmitted polarized light while the copy object is subjected to stresses in simulation of those to be applied to the real object. Areas of opacity and clarity appear in the copy object which can be related to stress concentrations. As a means of standardizing the perceived colour of objects, polarized light may be used in certain cases (4.3.10). Some notes on the uses of polarized light are given in Appendix V.

3.4.1i *Schlieren technique.* This is a method of test using polarized light which enables a photograph to be taken revealing differences in air temperature during heat testing of devices, showing concentrations of convected air and various temperature zones. Another method using heat-sensitive film which registers different colours for different temperatures produces similar results.

Chapter 4

# Lighting and colour

In this Chapter, before dealing with the obviously important technical matter concerning lighting and colour, the subject of colour in the factory is first examined (4.1). This topic is greatly neglected, and worthy of far greater attention than is generally accorded it. Correct colour treatment of factory interiors is second in importance only to the lighting, and is a major factor in its success or failure[16]. After this examination of the environmental aspects of lighting, the colour properties of light sources are discussed (4.2), to lay the foundation of explanations of how colour-matching and standardizing of colour is carried out (4.3). Finally, practical matters of factory layout and procedures are dealt with to show how the colour-matching techniques can be applied with accuracy (4.4).

## 4.1 Colour in the factory

**4.1.1** The environmental effects of colour in industrial interiors are important and far-reaching, for the general colours with which eyes are presented help us to form our psychological attitudes to the space we occupy. Well chosen colours, in combination with good lighting can make a useful contribution to the wellbeing of the occupants, stimulating morale and promoting productivity[17]. Modern practice tends to the use of large areas of white and grey, with little relief by strong colour; where such colour schemes are in areas having very well diffused lighting, the effect may be to create what is termed a 'bland field'. In a bland field, the eye finds no dominant feature upon which to focus, and some subjects find the effect most disturbing, to the extent that they may find it impossible to work happily in the room, and they may complain vociferously about the lighting. The adjustment of the colour scheme by introducing a small amount of strong colour brings about an immediate improvement in satisfaction with the interior (3.2.6c).

**4.1.2** We must distinguish between the illuminance provided in an interior (e.g. lux on the working plane) and the subjective sensation of *brightness* produced by the lighting. Brightness is produced by the reflection of light from a surface, and is related to the reflection factor of the surface and its colour. An interior painted in dark colours will reflect little light, and will appear gloomy even if the illuminance provided is generous. The *CIBS/IES Code*[5] gives guide-lines for devising luminance distributions in interiors which will be pleasing to the eye and tending to minimise glare (2.2.3). When choosing colours for an interior, the effect of reflectance factor on the brightness appearance should be considered, and it is quite likely that the lighting engineer designing the lighting scheme may recommend the adoption of certain preferred ranges of reflectance for the walls, ceiling, flooring and furniture (especially the bench-tops and table-tops). Reflectances are usually expressed as decimal fractions, e.g. 0.5 etc., and it is simple to relate these to the reflectances of paints if the *Munsell Colour Number* of the paint is known. (In a paint description, the Munsell Colour Number follows the group letter in the Munsell designation. Reputable paint manufacturers can provide this information about their standard colours). The reflectance and the Munsell value ($V$) are related thus:

$$\text{Reflectance} = \frac{V(V-1)}{100}$$

**4.1.3** In areas where fine colour tasks are done—e.g. colour-matching and colour inspection, there is an important restraint on choice of colours for decor in that they may affect the colour performance of subjects in the room (4.3). Degradation of ambient light by reflection from strongly coloured surfaces may also affect visual clarity (4.2).

**4.1.4** Colours to be used in the decor of an interior should be selected in the light of lamps identical with those to be used in that interior. The appearance of the paints may change significantly when viewed by the light of various illuminants. One of the many factors which may affect choice of colours for decor is the effect of *after-images*. Looking at a colour for long enough to become at least partly colour-adapted will result in after-images of the complimentary colour becoming visible when the gaze is transferred to a light-coloured neutral or white surface. This can be distracting, so if the task will consist of handling objects of a particular strong colour, then it could be beneficial if a 'visual rest area' for the eyes to rest on (e.g. the background of the task) could be of the complimentary colour thus neutralizing the experience of after-image. This is now general practice in operating theatres, the traditional white masks, gowns and drapes having been replaced with green ones; thus the reflected glare from the overhead lighting is minimised, and the surgeon is now not troubled with the green after-images he would sometimes see in the past as he transferred his gaze from the site of the surgical operation to some adjacent white surface.

**4.1.5** The effect of colour-adaptation should be noted in choosing decor colours. If, for example, a room is decorated with substantial areas of a 'warm' colour (pinks, reds), an occupant will soon become colour-adapted to the 'warm' end of the spectrum, with the result that if he passes from that area to an adjacent area decorated in neutral tones, he will judge that other area as being of 'cold' appearance until he becomes re-adapted to its colour tones.

**4.1.6** The colours of decor contribute to the visual performance of the occupants of an interior. By helping to form a pleasant and humane environment, the colour scheme will contribute to creating good morale and, indeed, to the promotion of good health of the occupants (1.2.1). It is not only for aesthetic reasons that it has been necessary to create a British Standard for colour co-ordination in buildings[18], for without some harmonization it would be difficult to find suitable coloured finishes for building components. There are a number of British Standards on colours for building components, including the cladding, and floorings[19]. Colours are alloted significance in a method of coding pipelines in factories[20] and a further series of colour significances is of great importance to safety in identifying informational, prohibitory and mandatory signs in factories (Table 3), while another series of colours signify voltages (Table 4). These matters are of importance on two counts: (a) colours of decor must not be confusable with significant colours (e.g. Tables 3 and 4); and (b) the light-sources used must have suitable colour-rendering to enable these colours to be readily recognized (1.1.6, 4.2).

**Table 3** Colour Code (BS Safety Colours)

| *Colour* | | | *Significance* |
|---|---|---|---|
| Red | 04E53 | 'Flame' | Fire equipment and alarms |
| Yellow | 08E51 | 'Gorse' | Where accidents are likely to occur |
| Green | 14E53 | 'Neptune' | Escape routes |
| Blue | 18E53 | 'Gentian' | Safety instructions |

**Table 4** Colour Code (BS 4343/CEE17 Voltage Colours)

| *Colour* | *Voltage* |
|---|---|
| Violet | 25 |
| White | 50 |
| Yellow | 110/130 |
| Blue | 220/240 |
| Red | 380/415 |
| Black | 500/750 |

## 4.2 Colour properties of light sources

**4.2.1** Vision is only possible in presence of light which enables us to form images on the retinas of the eyes[1], and the quality of that light affects how well we perceive our tasks and things about us. If the spectrum of the light-source in use does not contain certain energy in particular wave-bands, both the *colour-appearance* (4.2.2) and the *colour-rendering* (4.2.3) are affected, and with them the accuracy of our colour-perception.

**4.2.2** Colour-appearance of light-sources may be described as 'cool', 'intermediate' or 'warm'. 'Cool' colours are from the blue end of spectrum, 'warm' colours from the red end. The colour-appearance is the colour we perceive a light-source to be when we look directly at it, or when examining a white object in its light. Colour-appearance of two or more sources can be compared side-by-side, but once we enter a visual environment we become adapted to it and are less conscious of the colour-appearance.

**4.2.3** Colour-rendering is the property of the light from a source to reveal colours of objects. This may be tested within the limits of accuracy of the human eye by attempting to *match* colours in its light. Matching may be performed by mixing pigments in an attempt to reproduce a sample colour; or it may consist of trying to select two identical samples from a group of nearly similar ones. A match made under poor colour-rendering light will be revealed as poor match when the test is repeated under light similar to north-sky daylight. A true match holds good under other light-sources; a poor match will hold good only under certain light sources ('metamerism').

**4.2.4** *Colour temperature* is a method of defining the colour of a light-source. It is only an accurate description if the light-source emits a continuous spectrum (which most practical light-sources do not) but is a useful indication of light properties. The colour-temperature is given in degrees Kelvin; for example, there is considerable difference in colour-rendering and colour-appearance between a 'White 3500 K' fluorescent tube and one described as 'Daylight 4300 K', the latter (being of higher K) appearing much more blue in appearance in direct comparison with the latter. The higher the K value, the bluer the colour-appearance, and, for lamps of continuous spectrum, the better the colour-rendering (Figure 4.1).

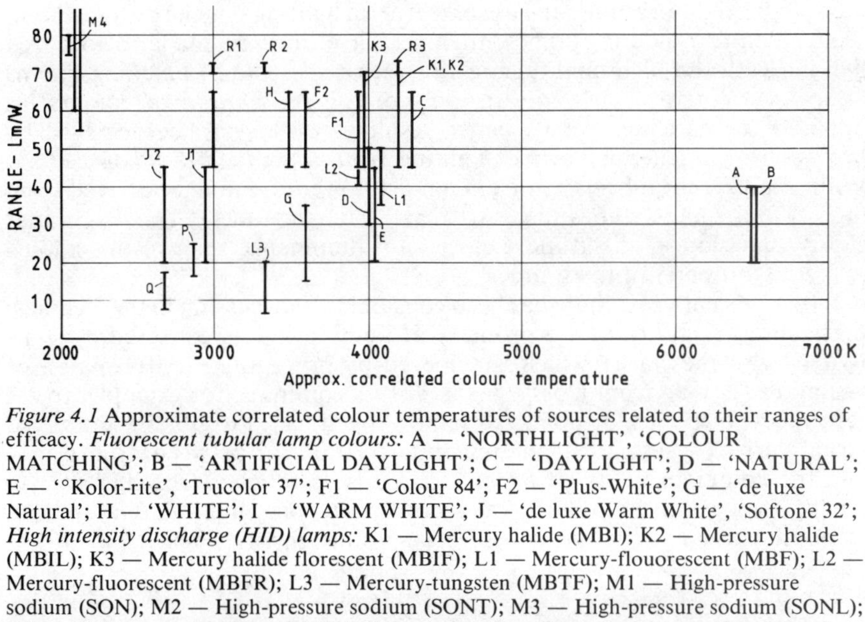


*Figure 4.1* Approximate correlated colour temperatures of sources related to their ranges of efficacy. *Fluorescent tubular lamp colours:* A — 'NORTHLIGHT', 'COLOUR MATCHING'; B — 'ARTIFICIAL DAYLIGHT'; C — 'DAYLIGHT'; D — 'NATURAL'; E — '°Kolor-rite', 'Trucolor 37'; F1 — 'Colour 84'; F2 — 'Plus-White'; G — 'de luxe Natural'; H — 'WHITE'; I — 'WARM WHITE'; J — 'de luxe Warm White', 'Softone 32'; *High intensity discharge (HID) lamps:* K1 — Mercury halide (MBI); K2 — Mercury halide (MBIL); K3 — Mercury halide florescent (MBIF); L1 — Mercury-flouorescent (MBF); L2 — Mercury-fluorescent (MBFR); L3 — Mercury-tungsten (MBTF); M1 — High-pressure sodium (SON); M2 — High-pressure sodium (SONT); M3 — High-pressure sodium (SONL); M4 — High-pressure sodium (SONR); *Other lamps:* P — Tungsten-halogen; Q — Tungsten-filament (GLS); R1 — 'Polylux 3000' tube; R2 — 'Polylux 3500' tube; R3 — 'Polylux 4000' tube. (*Data based on IES/CIBS Code*[5] *and manufacturers' information.)*

**4.2.5** *A colour-rendering index* (CRI) system has been devised, which is explained in the *CIBS/IES Code*[5]. Useful though it is to guide choices between possible types of lamps for a particular application, it is not very accurate, and two lamps having the same CRI are not necessarily interchangeable for the same critical colour duty[21]. Colour rendering index is represented by the symbol Ra.

**4.2.6** It is sound practice to use lamps of one type and colour in an area, and replacement lamps should be identically similar to those specified for the scheme. However, when seeking to obtain special effects, lighting designers use combinations of all kinds of lamps. A detailed summary of the available types of lamps and their colour properties is given in the *CIBS/IES Code*[5], and a selection of lamps suitable for industrial use is given in Appendix III. Low illuminances obtained from the 'cooler' fluorescent tubes and HID lamps tend to look dismal; for illuminances below around 300 lux, use light-sources of below 4000 K. Conversely, high illuminances from fluorescent tubes of 'warm' colour-appearance (viz low K) can be overpowering and give the impression of creating much heat; but lamps of better colour-rendering and of cooler appearance (viz high K) when used to produce illuminances of around 1000 to 3000 lux produce acceptable installations.

**4.2.7** Most factory lighting installations these days employ HID (high intensity discharge) lamps for mounting heights of 4 m or more. If the colour requirements are not too demanding, SON or MBF lamps are used; MBI lamps have rather better colour performance in this duty. Below around 4 m mounting height, the commonly used sources are fluorescent tubes (8.1). If there is a requirement for critical colour vision in a location where the mounting height precludes the use of a general lighting scheme of fluorescent 'better colour rendering' tubes, then recourse must be made to localized lighting with the high-quality lamps or the use of colour-matching booths (4.3, 4.4). Where a task consists only of the *recognition* of colours (as opposed to matching them), better results are likely to be achieved by provision of a generous level of lighting from a high-efficacy source (e.g. White fluorescent tubes) than from a much lower illuminance derived from high colour quality tubes (e.g. Artificial Daylight tubes). This is because clarity of vision is linked more closely to illuminance than to the colour-rendering property of the source.

**4.2.8** In recent years there has been considerable discussion in the technical press about *visual clarity*, a property of lighting that is confused by some writers with the clarity which is achieved by preventing the formation of veiling reflections from glossy surfaces. It is common, for example, to see references to the use of polarized light (Appendix V) to obtain 'improved visual clarity'. A simple explanation of the relationship between visual clarity and the colour property of a light-source is as follows. The human eye, having evolved over millions of years under solar light, has a spectral response curve matched to that of solar visible energy (Figure 4.2). It has

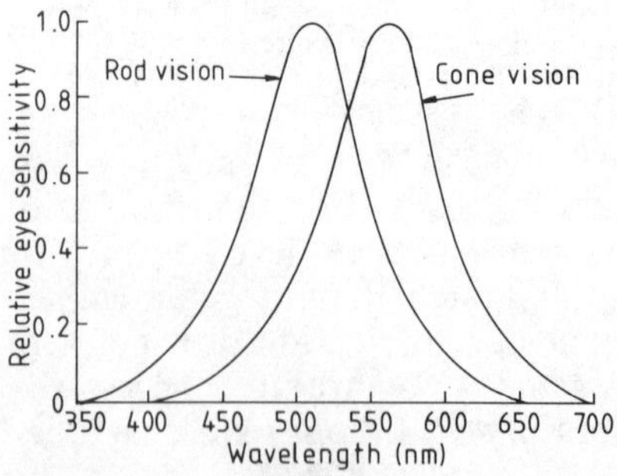


*Figure 4.2* The relative sensitivity of the human eye to wave-lengths of radiation, after Pritchard.[51] (The process of human vision is discussed in another book by the author.)[1]

long been the aim of lamp makers to produce a lamp in which the spectral power distribution followed closely to that of the human eye response curve. The technical difficulties in doing this are great, but a number of practical light-sources have been developed which, while having spectral power distribution envelopes of quite different contours to that of solar light, give satisfactory performance for many applications. Incidentally, lamp manufacturers publish such curves for their lamps, but it takes much expertise to interpret them and visualize the colour performance which may be associated with each shape of curve. It has long been known that other spectral distributions could give the sensation of white light, or of colours; for example, the combination of a light of orange colour (650 nm) and green colour (490 nm) present the eye with the colour appearance (say, when mixed by illuminating a white object) of pale blue—but, of course, without any rendering ability except for those specific orange and green colours (Figure 4.3). In recent years it has been discovered that the combination of

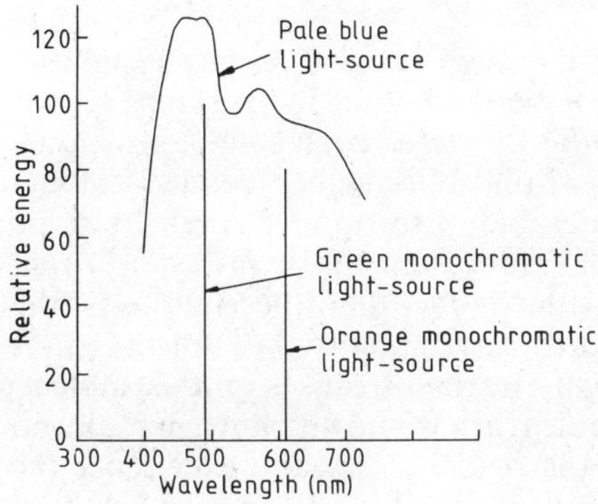


*Figure 4.3* The combination of the two monochromatic light-sources produces the same colour-appearance (but not the same colour-rendering) as the single pale-blue light-source. After Pritchard.[51]

only three specific colours, blue-violet, green and orange-red, could produce the sensation of a satisfactory white illuminant, and with fair colour-rendering too. Thus, it has been possible to develop tri-phosphor fluorescent tubes which combine high efficacy with good colour properties. Such tubes provide a colour-appearance which may be 'cool', 'intermediate' or 'warm' according to their formulation, and, although they tend to distort colours somewhat, many people actually prefer them to tubes having more or less conventional spectral characteristics. The colour-rendering may be measured and specified according to BS 1853[52], by 8-band or by 6-band analysis, or the CIE uniform chromaticity scale; such matters lie outside the scope of this book. What is important to grasp is that when people are shown interiors and objects lighted with these three-colour tubes, they express *preference* for them. They like the somewhat exaggerated colours; they actually prefer the effect of these tubes on human complexions, meat, vegetables, fruit, flowers and foliage. Leaves, for example, appear to be somewhat more blue-green; complexions seem pinker and healthier; a picture of the sky is rendered bluer than is the real sky we see. It is necessary to differentiate between *colour-rendering* and *colour-preference* (4.2.9).

**4.2.9** Colour-rendering is expressed by a Colour Rendering Index or CRI. Unfortunately, there is not always a direct correlation between the CRI and the colour-temperature of the sources, as may be shown by study of Table 5. Generations of lighting engineers and users have been used to the general proposition that "better colour-rendering sources are always of cooler colour-temperature'', but this is clearly no longer true. However, for non-critical applications of light-sources, e.g. for lighting interiors pleasantly, and for the recognition of colours (but not necessarily their accurate matching), then a Colour Preference Index (CPI) may be determined by a similar procedure to that employed for determining the CRI. Thus we have three colour factor measures for a light-source:

(a) The colour-temperature, expressed in degrees K, and giving a measure of coolness or warmth of the appearance of the source;

(b) The CRI or Ra value which expresses the ability of the light-source in matching colours (i.e. in identifying identical colours and differentiating between colours of differing properties and composition); and

(c) The CPI, which is entirely based on subjective preference, and may have no relation to either K or CRI.

It has been the aim of researchers to discover *why* subjects prefer the lighting effect of certain light-sources; it cannot be the colour temperature K, for tubes of warm, intermediate or cool K may be preferred; it cannot be colour-rendering alone, for, although the tri-phosphor tubes render colours attractively, there is often quite considerable distortion. It is felt by some workers that preference is based on something which is described as *visual clarity*, which seems to be associated with the fact that in certain tests, the light from these preferred lamps appear to be brighter to the subject than is indicated by the lightmeter. Despite many learned treatises on this subject, the author does not yet feel that the whole truth is known, and can make no recommendation apart from advising the reader to try and experience the new tube colours for himself under various conditions, and see what he prefers. But, a word of warning; where the colour properties of tubes are extolled, be certain that the tubes used for critical colour-matching are not those merely preferred. Tubes for critical colour work should be limited to only those which have been proved to avoid serious metamerism (false matching), namely, ARTIFICIAL DAYLIGHT, NORTHLIGHT and COLOUR MATCHING, the familiar 'British Standard Colours'.

**4.2.10** The lamp business is a highly competitive business. Each lampmaker vies with the others, trying to bring out a better lamp, a lamp that the customers will buy. Each makes claims as to the colour properties and the efficacies of his lamps, striving to survive in a tough marketplace. It was therefore perhaps to be expected that when certain lampmakers produced their three-colour/triphosphor tubes, some other manufacturer would come along with the offer of a 'continuous spectrum tube'. It happens to be a feature of the three-colour/triphosphor tubes that they tend to produce rather less ultraviolet emanation than do conventional tubes; so it was therefore again to be expected that the makers offering the 'continuous spectrum' tubes should make a feature of the fact that their tubes produced rather more u.v. emanation than conventional tubes. These matters, taken up and written about badly and mistakenly by certain journalists, created a Nine Days Wonder in the lighting world during 1980, with the appearance of a certain organization (its name is not recorded here—not wishing to appear to give it credence, nor give it publicity) whose aim has been declared to be to persuade the general public that 'Only continuous spectrum tubes with enhanced u.v. are conducive to good health', a statement of obvious untruth.

## 4.3 Colour-matching; standardizing

**4.3.1** Many misconceptions exist about colour-matching, and it would be possible to devote many thousands of words to explaining all that is known about this complex and fascinating subject. It must suffice here to review the essentials only, stressing that, for certain very accurate work, even the precautions and guidance given here may not be sufficient to achieve the very highest standards of accuracy and constancy. However the notes in this section, and those on practical techniques (4.4) should enable any of the commonly met industrial colour-matching tasks to be performed with ease.
**4.3.2** The first essential for accurate colour judgement is constancy of the illumination. Contrary to popular opinion, natural light is a poor illuminant for colour-matching purposes, for it continuously changes according to the season, the time of day and the weather. The spectral composition of daylight may not even be the same in two samples of light entering simultaneously through windows on opposite sides of a room. Traditionally, many critical colour tasks have only been done in north-lit rooms by daylight, e.g. hop grading, where at one time the grading was deferred annually until the crop samples could be valued under 'a natural cold north light, with the sky unobstructed by heavy clouds or fog'. The delay involved keeping samples while waiting for the 'ideal conditions' to undertake the valuation, (the 'wrong' light could considerably alter the appearance, and the value, of the sample), meaning that interest had to be paid on borrowed money until the grading had been done and trading could start. The assessment is now carried out entirely under electric lighting (using Colour Matching/Northlight tubes at an illuminance of 1500 lux) and the valuations can be completed several months earlier, with considerable cost-savings. It also seems that the consistent lighting conditions have resulted in more consistent valuations, so that appeals against valuations now seldom occur.
**4.3.3** Some examples of how the colour quality of the lighting affects accuracy of matching may be quoted. The fabrics buyer of an important London dress house said that under best quality electric lighting (Artificial Daylight tubes, 3000 lux) she could easily identify no less than twelve distinct shades of black, about twice as good as her performance under daylight. Since the provision of this standard of lighting, complaints about mismatched colours were rare, yet previously there had been a significant number of complaints. The complaints had often been that a single panel in a skirt, according to the customer, 'appeared to be a different colour at the dance, though it had looked all right at home,' a good example of mismatch that showed up under a different illuminant. Another case concerned a UK firm of printers who obtained an important order from a client in California. Knowing that the client was insistent on accurate matching of the inks to the samples, the printer took the precaution of matching them carefully under north-sky lighting conditions. Alas, the spectral quality of English daylight is very different to that under the blue skies of California, and he had to stand the cost of having the work rejected. Re-matching under electric lighting (Artificial Daylight tubes to BS 950, 3000 lux) resulted in a match which the client approved without demur. In this case, the failure to get a match first time was probably partly due to the fact that the inks or the dressing in the paper fluoresced somewhat under the u.v. component of light. Many such

examples can be quoted, dealing with products such as bank-notes, stamps and securities; foodstuffs, fabrics and paints; plastic coatings on sheet steel, and the colour of anodised finishes on aluminium.

**4.3.4** The provision of correct conditions for accurate colour-matching involves more than just *choosing the right tube* (4.3.5). It is also essential to *provide the correct illuminance* (4.3.6), to ensure there is adequate time for the inspector's personal *colour-adaptation* (4.3.7), and steps must be taken to avoid errors due to the phenomenom of *colour reduction* (4.3.8). Only by attention to all these matters will it be possible to get a good level of *standardization* (4.3.9). Finally, the colour-matching system has to be integrated into factory practice by good *applications techniques* (4.4).

**4.3.5** Choice of the best light-source for colour work is vital, and for all really critical work the choice will be a fluorescent tubular lamp. These lamps are low-pressure linear light-sources, which have differences in colour-appearance and colour-rendering due to their being provided with different phosphor coatings. Lamps of higher colour quality tend to be of lower efficacy. These lamps will operate in all positions. The characteristics of the colours of tubes used for colour discriminatory work are given in Table 5.

**4.3.6** As regards the illuminance to be provided for colour work, the recommendations of the *CIBS/IES Code*[5] are a sound guide, and the illuminances quoted there should be regarded as minima. The eye cannot achieve its potential performance in discriminating colours unless it is light-adapted, and for practical purposes an illuminance of around 1000 lux is the minimum at which any reliable colour work can be done. There is virtually no upper limit to the illuminance at which a subject may find visual comfort, for we can adapt to illuminances as provided by the sun of 80 000 to 100 000 lux. In practical tests, a large proportion of subjects judged illuminance to be satisfactory around 2000 lux (Figure 4.4) at which level the

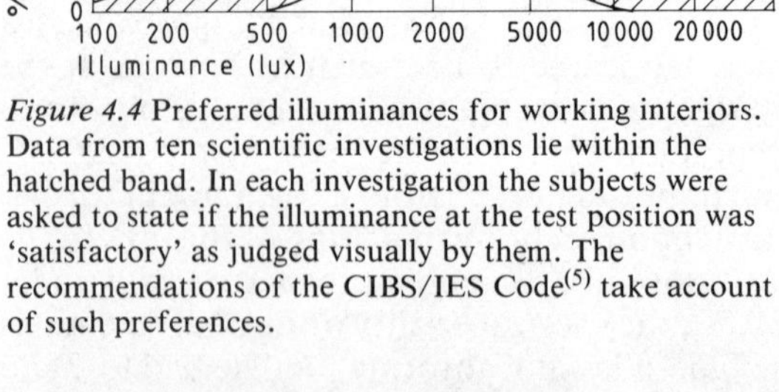


*Figure 4.4* Preferred illuminances for working interiors. Data from ten scientific investigations lie within the hatched band. In each investigation the subjects were asked to state if the illuminance at the test position was 'satisfactory' as judged visually by them. The recommendations of the CIBS/IES Code[5] take account of such preferences.

eyes will be fully light-adapted very quickly. It is thought that the satisfaction of the test subjects partly stemmed from the greater clarity of vision that occurs at this order of magnitude of illuminance when the subject's eyes register colours fully. It is also believed that the tail-off of those satisfied at illuminances above 2000 to 5000 lux may have been at least partly due to

**Table 5** Colours of fluorescent tubular lamps. This table should be read with the accompanying notes

| *Tube name (a)* | *Circuit efficacy (including control gear loss) (b)* | *Colour appearance (c)* | *Colour temperature K (d)* | *Colour rendering index Ra (e)* |
|---|---|---|---|---|
| ARTIFICIAL DAYLIGHT | 27 | Cool | 6500 | 95 |
| NORTHLIGHT } COLOUR MATCHING } | 35/38 | do | 6500 | 94 |
| Maxilux daylight | 68 | do | 6300 | 85 |
| DAYLIGHT } Cool white } | 58/61 | do | 4300 | 67 |
| Low-watt cool white | 65 | do | 4300 | 67 |
| Polylux 4000 | 69/70 | do | 4100 | 85 |
| Maxilux white | 72 | do | 4100 | do |
| Trucolour 37 | 41 | Intermediate | 4000 | 98 |
| °Kolor-rite | 39/41 | do | 4000 | 92 |
| NATURAL | 44/46 | do | 4000 | 85 |
| Colour 84 | 62 | do | 4000 | 98 |
| Plus white | 58/62 | do | 3600 | 74 |
| De Luxe Natural | 32/36 | do | 3600 | 92 |
| Polylux 3500 | 69/70 | do | 3400 | 85 |
| WHITE | 62/66 | do | 3400 | 56 |
| Low-watt white | 69 | do | 3400 | 56 |
| Maxilux warm white | 72 | Warm | 3000 | 85 |
| Polylux 3000 | 69/70 | do | 3000 | 85 |
| Colour 83 | 69 | do | 3000 | 85 |
| DE LUXE WARM WHITE | 42/44 | do | 3000 | 79 |
| WARM WHITE | 60/64 | do | 3000 | 54 |
| Low-watt warm white | 65 | do | 3000 | 54 |
| Softone 32 | 44 | do | 2900 | 85 |

(a) Names in capital letters denote British Standard Colours. Lamps grouped together in the table are not necessarily identical in performance.
(b) The circuit efficacy figures are typical, and have been compiled by calculating the lumens per watt on the basis of the Lighting Design Lumens of lamps of 80 W, 65 W or 58 W divided by the approximate total circuit watts in each case. The figures quoted may differ slightly from the manufacturers' published figures, and may be different for other powers of tubes.
(c) The colour appearance column is based on data in a similar kind of table in the CIBS/IES Code.[5]
(d) The values of K denote the approximate correlated colour temperatures in Kelvin, a measurement that strictly should only be applied to spectral distributions closely approximating to that for a 'black body radiator'. But this is a system that has been widely used in the past, and the lamps are arranged in the table in descending magnitudes of the K value.
(e) Within each group of lamps of the same value of K, the lamps are arranged in the table in descending order of the value of their Colour Rendering Index (Ra). The CIE Colour Rendering Groups are:
Group 1: Very good. Ra = 85 to 100 (Suitable for colour-matching)
Group 2: Good. Ra = 70 to 84 (Suitable for colour-recognition)
Group 3: Sufficiently acceptable for use in general working interiors. Ra = 50 to 69.
The steps on the Ra scale are not uniform, but tend to give rather higher value to lamps of lower K, viz over the range of lamps where the lamp output is fairly close to the characteristic of a black body radiator.

increasing discomfort from glare under the test conditions. Trials of lighting methods under industrial conditions do not usually produce convincing results (because of distractions to the experimenters and difficulties of controlling conditions), but trials of lighting for colour vision often leave the experimenters with the conviction that somewhat higher illuminances than those recommended in the *CIBS/IES Code*[5] do, in fact, bring about better performance in colour work. If the task illuminance is only required over a relatively small area, the cost of providing an illuminance of the order of 3000 to 5000 lux is not great, and is an experiment well worth trying if the colour task is critical. Table 6 quotes some examples of illuminances recommended in the *Code* for visual tasks involving critical colour discrimination.

**Table 6** Examples of recommended illuminance for colour tasks

| *Task or location* | *Standard Service Illuminance (lux) (See Note)* |
|---|---|
| Bakeries | |
| Decorating, icing | 500 |
| Boot and shoe factories | |
| Cutting tables and presses | 1000 |
| Carpets | |
| Inspection | 1000 |
| Dye works | |
| Dyehouse labs, dyers' offices | 1000 |
| Final examination | 1500 |
| Furniture factories | |
| Veneer sorting and preparation | 1000 |
| Leather working | |
| Grading, matching | 1500 |
| Paint works | |
| Colour matching | 1000 |
| Printing works | |
| Printed sheet, inspection, precision proofing, retouching, etching | 1000 |
| Colour reproduction and printing inspection — colour and registration | 1500 |

These examples are taken from CIBS/IES Code.[5]
The illuminance quoted are for the bench or working plane; in some cases, additional directional local lighting is also needed.

**4.3.7** Adaptation from one illuminance to another takes an appreciable time. If an inspector moves from an area of the factory lighted to a few hundred lux and enters an inspection area or booth lighted to several thousand lux, he must allow time for his eyes to adapt to the higher illuminance before starting to try and exercise critical colour judgment. The time for adaptation will vary from subject to subject according to personal characteristics and age, and on the first and second illuminances (Figure 4.5). The importance of this lies in the fact that the subject can only achieve his potential performance in colour discrimination when well advanced in to light-adaption (termed photopic vision). Adaptation to around 1000 lux probably represents the minimum acceptable standard of colour performance for industrial colour-matching work, with detectable improvements in some subjects (particularly older ones) up to 5000 or even 10 000 lux if glare is controlled and if adequate time is allowed for adaptation.

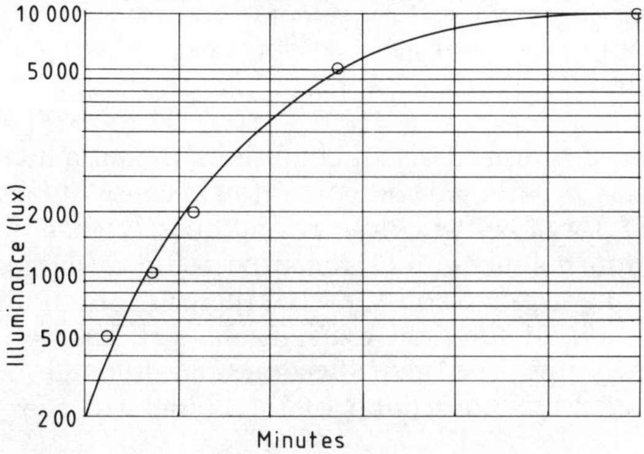


*Figure 4.5* Time to adapt to a higher illuminance. The subject can only achieve his potential performance in colour discrimination when well advanced into light-adaptation, and adaptation to 1000 lux probably represents the minimum standard for industrial colour matching. The time to adapt from a lower to higher illuminance (results of the author's own experiments) are indicated by the curve, viz to adapt from 1000 lux to 5000 lux takes about $1\frac{3}{4}$ minutes, and from 500 lux to 5000 lux about $2\frac{1}{4}$ minutes.

**4.3.8** A simple experiment will demonstrate a factor that complicates colour-matching work, namely the phenomenon of *colour reduction.* On a sheet of strongly coloured material, say of bright red, place a small piece of neutral-grey paper or felt measuring about 10 mm square. Let the sheet be illuminated well, say at 500 lux or more, and most observers will be of the opinion that the grey sample is of a green colour. Repeat the experiment, but this time placing the small grey sample on a green background sheet, and this time the observers may conclude that the grey sample is of a pinkish colour. These results are due to the eye tending to 'see' the complimentary colour to that to which it is adapted, viz an *after image* is superimposed on the grey sample and the illusion of it being coloured is created. The aberration can be even more confusing; if the small grey sample is replaced with a pale blue sample, on the red background it will look green, while on the green background it will appear as a pale tint of purple, and if placed on a further background of darker blue it can actually appear to be white. Quite obviously, such serious errors must be eliminated from practical colour-matching work. The experiments just described are extreme examples, but some effects similar to these will be met in practice if the background to the visual task is so strongly coloured as to bring about a degree of adaptation to that colour. For this reason, in areas where fine colour work is performed, e.g. matching of colours and inspection of coloured materials, colour printing on paper and fabrics etc, care has to be taken to restrain the strength of colours in the decor to avoid degrading the colour-quality of light reaching the task (4.3). It is good practice to restrict the colours used for objects and decor in such areas to very light colours, not stronger than Munsell Chroma No 1. (Attempts to avoid colour reduction by providing an all-white environment may result in the occupiers suffering from after-images, and these can produce a most uncomfortable sensation, in some cases verging on visual hallucination. A totally colourless interior is an unnatural environment (3.2.6c)).

**4.3.9** Efforts to create standardization of visual conditions so that a high level of colour performance may be attained sometimes fail because of lack of attention to small details. As well as the colour reduction effect due to inadvertent adaptation to random colours (4.3.8), the inspector may find his accuracy is poor in a room having strong colours, even if these are out of his line of view; this is because some of the light reaching his task has been reflected from those surfaces, and the light becomes adulterated or 'degraded' because of the preferential reflection of the colours of those surfaces.

In one case seen by the author, results on matching fine colour prints were below the required standard; investigation showed that the light on the task (1000 lux from 'Artifical Daylight' tubes to BS 950 Part I) was being degraded by reflection from the pale green plastic floor tiles in the room. In a convincing experiment, the floor was temporarily covered with sheets of white paper, when the quality of colour judgment needed immediately became possible. Incidentally, raising the reflectance of the floor greatly lightened the appearance of the walls and ceiling, and raised the illuminance on the bench tops by about 10 per cent.

In another case, the standard of accuracy was not consistent in the inspection of plastic colour samples, with the mysterious effect of getting differing results from the same samples on different days or by different inspectors. This turned out to be due to specular reflections of the luminaires appearing on the surfaces of those colour samples which were more highly polished than others. Again, a simple experiment proved the theory right; selected luminaires were provided with temporary diffusers in the form of sheets of tracing paper, and the diffusion removed the inconsistency in the results immediately. Later, the luminaires were fitted with suitable permanent diffusers. Another cause of inconsistency observed has been variance in the colour-rendering of Northlight/Colour Matching fluorescent tubes with variation in temperature. In a factory, the user had tried to standardize conditions by stipulating that the tubes should be switched on 30 minutes before work commenced, so that the tubes could reach their normal operating temperature. Costly mistakes (in the matching of screen-printed silks) later prompted an investigation, when it was discovered that a pocket of still hot air formed close to the ceiling after some hours of use of the lighting, and the tube wall temperatures rose some 10°C with a discernible colour shift. Improvement of ventilation and better control of temperature at the luminaires (which were lowered 300 mm) overcame the problem completely.

**4.3.10** A complication of devising lighting to suit difficult colour tasks is that many substances exhibit *dichromaticity,* i.e. that they exhibit different colour characteristics under different conditions of lighting and viewing. This may take place in either the transmitted or the reflected mode. Examples of the transmitted mode are many; for example, the metal gold is of its familiar colour when viewed by direct light, but a thin sheet of gold foil transmits only green light, and thus appears to be green by light transmitted through it. Liquids may exhibit this phenomenon too; bottles of chemical solutions (and some wines) may appear to change colour when viewed against a bright light. There are also many examples of the reflected mode of dichromaticity; for example, silverside of beef looks mainly red when viewed normal to the surface of a cut, but viewed at a glancing angle the 'silver' aspect is more noticeable. Similarly with certain printed papers, particularly material that has been lithographed, e.g. some postage stamps which appear to change colour when viewed normal to their surface or at a flat angle. When the surface of an object which is semi-specular is examined, at normal angles its true colour dominates; but at flatter angles the appearance may be modified by colours reflected in it from other objects or from light-sources, and this effect can be negated by the use of polarized light (3.4.1(h)).

**4.3.11** The perceived colours of objects may change due to *fluorescence* under ultraviolet radiation from daylight or u.v.-rich light-sources. This is particularly confusing when one is trying to match materials of the same nominal colour but which contain chemically different dyestuffs as may occur in the paper, textiles and carpet industries. The effect also occurs with many foodstuffs (19.1).

## 4.4 Colour-matching techniques

**4.4.1** Apart from the mixing of paints, inks and pigments (where the standard formulation may need to be adjusted to take care of tolerances in the chromatic characteristics of the constituents), most industrial colour-matching tasks involve either *side-by-side matching* (4.4.2) or *differential matching* (4.4.3). Differences in colour arise from many possible causes, such as variation in paint thickness over a substrate; differences in browning of baked foodstuffs; differences in freshness or quality of meat; tolerances in formulation of various products which consist of mixed constituents; variations in processes such as plating, buffing, anodising and dying of metals; differences in production or dyeing of fabrics etc.

**4.4.2** In side-by-side matching, the sample under test is compared with the control sample. Good matching cannot be done if the inspector has to move his gaze very far between the two samples, particularly if on its way between the two visual objects the gaze has to pass over any other colour or something of different brightness. Ideally, the two samples to be compared should be contiguous and capable of being seen simultaneously under identical lighting conditions and at identically similar distances and angles. Where it is physically impossible to achieve this, a smaller control sample may be laid on a larger test sample or *vice versa*. It is essential that both samples are of the same degree of specularity, and that the control sample does not become damaged or dirty in use. It is vital to the operation that the control sample should have been standardized, e.g. inspected by at least two inspectors independently, and that the matching is performed under exactly the same lighting conditions as those under which the control sample was standardized and approved.

An example of a particular difficulty in control of colour being overcome concerns the browning of meat pies. An important feature of bakery products is the colour of pastry and flour goods; a pale pie is unappetising, a dark pie is suspected by the shopper of having been 'freshened up' in the oven. Just the right colour of pastry makes the product look tasty. Because of minor variations in materials and oven temperature and speed, one bakery concern checked every batch of pies against a 'perfect sample'; unfortunately, in the heat of the bakery, the 'perfect sample' soon deteriorated and changed colour. Therefore, a firm of model-makers were employed to fashion a replica of it as the control sample. The control sample had to be accurately coloured against the 'perfect sample', and checked under the same kind of illumination as that used for the monitoring of real pies, in this case under 'Artificial Daylight' tubes.

An ingenious method of getting a conjoined image of two samples of fabrics so they could be closely examined was developed by a research association, in which mirrors were used to bring two virtual images side by side. In another application, the control sample was placed above the test sample and facing downward; a small mirror placed on the test sample produced an image of the test sample right in the mass of the test sample. (These mirror techniques will work if a high quality mirror is used, preferably a front-silvered one (3.3.5b)).

**4.4.3** Differential matching is a sophisticated method of matching that lends itself to accurate quality-control, and, with the aid of a computer, the variance of performance of individual inspectors or groups of inspectors can be monitored. The method involves presenting the inspector with four samples: (a) the control sample, (b) the sample under test, (c) a 'dummy' sample containing a slight deliberate variance of hue, chroma or greyness from the control sample, and (d) another dummy sample containing a different slight deliberate variance. Only the controlling inspector knows which sample is which. One or more inspectors now examine the four samples, and have to attempt to identify them. If they can find two identical samples, this is recorded, or failing that, they try to decide which is which. The results recorded over a period of time for different groups of four samples as inspected by different inspectors can be analysed to find out a great deal of useful information, e.g. the minimum perceptible colour differences which can be detected by each inspector; the percentage variance of various colour components; variation with time of day, day of week etc.; constancy of performance of individual inspectors; the effect of slight environmental changes, including changes to the lighting. With a suitable programme, the data fed to it through a carefully designed answer form, (which may include ranking of judged colour attributes by the inspectors) an advanced computer can 'learn' about the inspectors, and be programmed either to weight their reports according to their past record, or to directly or indirectly vary the colour constituents of the production to bring the product closer to perfection.

**4.4.4** Before appointment, inspectors should be asked to submit to vision screening (3.1.5) which will include appropriate testing for colour vision. Testing should be repeated at two-yearly intervals, particularly for older testers. Annual testing might be advisable for inspectors over the age of 50. Colour discrimination ability tends to decline with age, due to the gradually yellowing that occurs in the cornea and crystalline lens of the eye, but this is to some extent compensated for by experience. The experience factor is important, and should not be dismissed; surgeons have been known to perform at extreme standards of accuracy in normal surgery and microsurgery, exercising a high standard of colour judgment, yet doing this work under tungsten-filament lamps which have a poor colour rendering property (see Table 5). On change to modern light sources of better colour-rendering, these surgeons have to undergo a period of re-learning before they regain their former ability. Similar effects may be found in industry; e.g. on change to an improved lighting system, there may at first be complaints that 'the new lighting is not as good as the old', but, after a period of re-learning, the inspectors will find the new lighting actually makes the job easier, and the quality of their work may improve.

**4.4.5** The performance of an inspector on fine colour judgments will not be constant. His accuracy will decline a little through the shift due to fatigue; his accuracy will decline a little more each day through the week, the onset of fatigue occurring earlier each day. The value of his work will be greatest on the first day of the week after his break, and least on the last day of the week when accumulation of fatigue is greatest. A short day, with frequent breaks, is required for high performance. Various workable patterns of hours have been employed by factories, including switching inspectors between colour inspection and other duties on alternate days or half-days, in order to give their eyes a rest. It is probably simple fatigue, not the use of the eyes, that causes colour judgment deterioration. Testing of individual inspectors' performance by the differential matching technique (4.4.3) gives rise to the belief that performance is worse the day after imbibing large quantities of alcohol, and that deterioration of performance after taking even modest quantities of drink persists for many hours. A heavy meal, particularly one with meat (which stimulates the bile) will cause temporary deterioration of colour judgment. Older inspectors, whose colour judgment appears to be gradually deteriorating, may find their skill restored by their being provided with higher illuminance, say 50 to 100 per cent more than needed by younger workers[4].

**4.4.7** Arrangements for special lighting for accurate colour inspection will vary according to whether a central inspection department undertakes the work, or if inspection is at points dispersed around the factory (3.2.4). Because of the relatively low efficacy of the lamps of good colour-rendering, general lighting of whole departments to the required illuminance for fine colour work may not be economic. There are three main ways in which this problem can be tackled: by providing general lighting from the high quality tubes but at a lower illuminance than that needed for inspection, plus local or localized lighting from the same sources at work-stations (4.4.8); or providing general lighting at a generous level but lamped with high-efficacy tubes, and providing inspection lighting at the work-stations from high colour quality tubes (4.4.9); or recourse to use of inspection booths, the general lighting being either from high efficacy or high colour quality tubes (4.4.10).

**4.4.8** Discussing the first of the three alternative techniques outlined in paragraph 4.4.7, consider the arrangement where the general lighting is provided by high colour quality tubes but at a lower illuminance than that needed for inspection, plus local or localized lighting from the same kind of sources at the work-stations. This is a workable method, but has the disadvantage that the general appearance of the interior may be unsatisfactory (4.2.6). Further the difference in illuminance between the general lighting and the task lighting must not be too great (2.2.2) or discomfort will be caused, and the adaptation time will be inconveniently long (4.3.7).

**4.4.9** Discussing the second of the three alternative techniques outlined in paragraph 4.4.7, consider the arrangement where there is provided a generous level of general lighting from high-efficacy tubes, with better colour-rendering tubes to produce inspection lighting at the work-stations. This is a workable method, but some form of screen has to be employed to prevent the light from the high-efficacy tubes from 'diluting' the light from the better colour-rendering tubes at the work-stations. The scheme will work best if the two types of lamps have approximately the same colour-appearance, which will reduce the degree of colour adaptation required on movements between the general area and the work-stations. If the general lighting is of the same order of illuminance as that at the work-stations, this will reduce the time needed for adaption to the work-station illuminance (4.3.7).

**4.4.10** Discussing the third of the three alternative techniques outlined in paragraph 4.4.7, consider the arrangement where inspection booths are used, with the general lighting being either of high-efficacy or high quality tubes. This is a very commonly employed method, but it cannot produce good results if adequate time for adaptation on entering the booths is not allowed before work commences (4.3.7). The system is generally similar to that discussed in paragraph 4.4.9, but with a greater degree of environmental enclosure about the work-stations. It is this enclosure which may cause problems. Some subjects suffer a degree of claustrophobia, and many people find it unpleasant when two or more persons must cram into a small booth for conference. The problem of avoiding excessive contrast between the illuminances within and without the booths (2.2.2) must be dealt with in design. A compromise is to employ a 'head and shoulders booth' construction as described by Bellchambers and Phillipson(11) (Figure 3.5), where the problems of ventilation are not so acute. Note that the fluorescent tubes in an inspection booth that is not adequately ventilated may overheat, with a resultant change in colour properties (4.3.9).

**4.4.11** The use of light-boxes (3.2.8) can be an aid to various kinds of colour work for examination of, or work on transparent or translucent objects and sheets. The use of a fan to cool the interior of the light-box will in some cases be necessary to prevent colour-shift of the output of fluorescent tubes which will occur if they overheat. The colour of the tubes inside the light-box should be the same as that used in the general lighting. For finest colour-transparency work, it is not possible to replace one tube that fails in a light-box, for this will show a band of high luminance (and possibly slightly different colour) on the diffusing light-box surface. Either all tubes must be replaced as a batch, or one or two extra tubes should be operated external to the light-box but controlled by the same switch, so these tubes have the same operational life as those within the light-box; these extra tubes are then used for replacement of those within the light-box on their failure. Although the rated life of the tubes may be stated by the manufacturer as 7500 h, for fine colour work replacement at 5000 hours—or even at 3000 h may be considered necessary to get the highest standard of colour constancy. As for other tubes used in colour work, those in a light-box should be switched on 30 minutes before the work starts, to allow time for stabilization of temperature and thus of colour.

Chapter 5

# Lighting and safety

The provision of good lighting, 'sufficient and suitable' for the environment and the tasks in the workplace—and in places to which visitors or the public may be admitted—is a legal requirement (see Appendix I). The subject may be examined under topics such as *glare, perception, fatigue* and *colour* (5.1), and the special topics such as *adventitious light* (5.2) and *discontinuous light* (5.3) must also be understood if one is to design lighting that will have minimum risk of producing visual error (5.4).

## 5.1 Accident causation

**5.1.1** It has long been recognised that the state of the lighting in a workplace has important effects on the frequency and severity of accidents, both those which cause personal injury and those which merely result in damage. For every reported or recorded accident there are probably several 'near-misses'; people who work in places where accidents frequently occur tend to work a little slower as a precaution, so there is probably some loss of productivity throughout the establishment. Accidents are costly; they not only affect the injured person, but involve other workers and senior staff who have to deal with the matter[(3)]. The drama of a personal injury can affect the output of those nearby for days, and a fatality in the works will depress productivity for the whole organisation for a long period. These considerations are additional to the humanitarian necessity of protecting people from injuries, as well as the need to comply with requirements of the law (Appendix I). There is no doubt that lighting does have an effect on the probability of an accident; but it is to be regretted that the form upon which employers must report industrial injuries does not require the state of the lighting at the scene to be recorded. Were this done, valuable statistics would be available, as has happened in the reporting of road accidents, where the forms completed by the police always note the lighting conditions at the scene of the incident.

**5.1.2** There is evidence that poor lighting might cause accidents in two separate ways, (1) by the psychological effect on the occupants in a workplace that is gloomy, as well as the fact that working in a poor light tends to make one easily fatigued and irritable—both factors which might predispose to some act leading to an accident,—and (2) where the lighting is so defective as to contribute directly to accident causation. The link between frequency of accidents and the lighting conditions has been a subject for study and observation for many years[(12)], and one continues to note the tragic toll of injuries and deaths that are reported[13][14]. Ill-lit premises soon become neglected and dirty; untidiness and the accumulation of rubbish and floor obstructions set the scene for accidents. Bright premises tend to be kept clean and tidy, and such 'good housekeeping' promotes safety and reduces fire risks too. Quite apart from these 'social' effects on behaviour, lighting may be the critical factor in accident causation, leading either to *failure to perceive* (5.1.3) or *incorrect perception* (5.1.4).

**5.1.3** Failure to perceive a danger may be caused by insufficient illuminance on the task or its surroundings, or by glare which handicaps vision. Dense shadows may lead to tripping or falling accidents. Poor lighting may prevent the subject seeing a projection upon which he may impale himself, or he may not see a stationary crane-hook and walk into it. Shadows and glare may prevent the subject noticing a gap between the tail-board of a lorry and the edge of a loading-bay. Over-spacing of luminaires will produce patchy illuminance and increase dangers, as will failed lamps or dirty luminaires. Under these conditions the subject simply may not see the danger until it is too late to save himself or a workmate from injury.

**5.1.4** Incorrect perception covers optical illusions which sometimes lead to mistakes about depths or distances, particularly when the light flow is from an unusual direction (1.1.5). Flickering lights, and light seen intermittently (for example through the spokes of a wheel) can produce some curious and highly dangerous errors of perception (5.3), while light which unexpectedly is extinguished can leave the subject exposed to danger (5.2, Chapter 6).

**5.1.5** Despite the provision of guards, many accidents occur because subjects put their hands into dangerous places, often because they try to gain information manually that is not available visually (3.2.6a). When the paper industry in UK enhanced the lighting at the reel-up in paper-mills, there was a significant reduction in personal accidents, for the better lighting enabled the operatives to *see* the tightness of winding without 'patting the reel' and placing their hands in danger (19.2.3). The lighting should enable the workers to get the information they need about their tasks without exposing themselves to danger, i.e. to *'use eyes, not fingers'*.

**5.1.5** It would be ironic as well as tragic if lighting intended to make the workplace safe was itself a source of danger. Considerable increase in safety will be obtained by the use of reduced-voltage for portable lighting, particularly that to be used in 'earthy' conditions (14.3); if 'temporary lighting' is properly installed and used (8.4); and if suitable portable or mobile lighting is available when it is required, so avoiding need for dangerous improvisations (Chapter 14).

**5.1.7** A possible source of accidents is the effect of moving lights, for example, down-lights fitted on overhead gantry-cranes in buildings. Persons working below can become confused in their orientation or balance when they find the shadows about them are moving[1]. For this reason the use of moving overhead lights should be avoided.

**5.1.8** Defective vision, as well as defective lighting, can lead to accidents, and the practice of regular eye-examinations or vision screening is recommended (3.1.5). Even for persons having nominally normal sight, there may be unsuspected dangers in proximity to fast-moving machinery if they suffer a defect of binocular vision which could be detected by vision screening (3.1.6).

**5.1.9** In the presence of glare, the 'seeing value' of the lighting is diminished; it is as though the illuminance was lower than that registered by a lightmeter. The dazzling effect can prevent the subject seeing danger, while the mental confusion glare can produce may cause him to fail to take protective or avoiding measures in time. Small points of glaring light can have a hypnotic effect, dazing and confusing the subject and making him more accident-prone[(12)]. The control of glare from luminaires and from adventitious sources (5.2) is thus seen to be of high importance in the prevention of industrial accidents.

**5.1.10** A form of personal injury that is almost entirely preventable is that due to sparks or foreign bodies entering the eye. Every employer has a clear duty under the law to cause persons exposed to this danger to wear face and eye protection[(23)]. Where workers are required to wear eye protection continuously, it is sometimes found they are reluctant to do so if the illuminance in the workplace is very low; it is good practice therefore to increase the illuminance in such workplaces one 'step' above the final service illuminance as determined by reference to the Flow Chart in the *Code*[(5)] (Figure II.2). Where it is impossible to prevent dust settling on the eye protection device, as well as ensuring that the device is cleaned as frequently as is practicable, it will also be found helpful if the worker wears a peaked work-cap, a peaked bump-cap or safety helmet, or a dark-coloured peak-type eye-shield. Such devices will reduce the brightness of any particles adhering to the eye protection device, and make the wearing of it more acceptable to the worker.

## 5.2 Glare and adventitious light

**5.2.1** Light which comes to the eye in an uncontrolled fashion, and which does not contribute to effective vision, is termed *adventitious light.* Frequently, adventitious light is a source of discomforting or disabling *glare* (1.1.7). Glare may also be caused by ill-designed lighting installations, though the following of established design procedures can limit the discomfort glare due to general lighting systems to an acceptable level (16.3).

**5.2.2** Examples of adventitious light include shafts of direct sunshine entering a building, and so increasing the range of luminances as to seriously handicap vision (1.1.7). Light from hot processes or furnaces can intermittently swamp the effect of the normal lighting. In a study of lighting conditions in one drop-forge, it was found that the amount of light available for movement depended entirely on the chance of how much light was being produced by the process at that moment—an intolerably dangerous practice[(4)].

**5.2.3** In locations where luminous processes occur (e.g. glass industry, foundries, drop forges, hot metal extruding) the general lighting should be set at an illuminance that will ensure (a) that there is adequate illuminance for safety in the absence of the adventitious light, and (b) if adventitious light cannot be controlled, then when it occurs the general illuminance is sufficient to avoid danger from disabling glare from the adventitious light (19.4.1). It will be found that the provision of light-coloured reflecting surfaces to walls and ceilings will greatly ameliorate the problem, but the maintenance of light-coloured surfaces in industrial areas may be difficult or costly.

**5.2.4** A common, and completely avoidable, cause of danger is *flash* from welding. Welding areas should be provided with fixed or portable screens to protect the eyes of anyone nearby or passing through the department and who is not wearing protective goggles. The screens not only give physical protection from sparks and metal splash, but also protect more distant persons from the adventitious light. It has been observed that where the general lighting in a welding shop is of poor illuminance, the welders often touch the electrode off to make a flash for the purpose of orientating the rod to the task, their problem being that once the hood is over their face, they cannot see the rod tip until the current is flowing. The provision of a focusing spotlight to put a small patch of high illuminance on the task area (say 1000 to 2000 lux) greatly aids the task and leads to better quality welding. (Directly looking at the welding flash can cause a painful and dangerous injury to the eye (retinitis) caused by exposure of the eye to potent wavelengths of ultraviolet radiation. Protective goggles should be worn at all times in welding shops).

## 5.3 Discontinuous light

**5.3.1** By the term 'discontinuous light' is meant light that is not of a steady quantity, but which varies in amplitude or is intermittent in nature. All light sources which operate from alternating current (a.c.) have some degree of rippling or continuous variation in light output at a frequency of twice that of the supply, e.g. at 100 and 120 Hz on 50 and 60 Hz supplies respectively. Some special light sources intentionally generate intermittent light, e.g. stroboscopic flash lamps. In some industrial situations, light that is substantially of a steady nature may be broken into discontinuous light by being viewed through the spokes of a rotating wheel, or seen through reciprocating machinery. Intermittent light can also be generated by the rhythmic movement of mirror-like surfaces which will reflect light, e.g. light being reflected from a polished surface on a vibrating machine.

**5.3.2** As regards light which is pulsating in a regular manner, it may or may not have a wave-form similar to a succession of half-waves of sinusoidal shape; more commonly the half-waves are overlapped so that the light is of a steady quantity with a ripple of amplitude. The ratio of the peak amplitude to the root-mean-square value of the wave is termed the 'form factor'. The form factor for steady light is thus 1, while that for a light which varied between its maximum and its rms value in a ratio of 2:1 would be 0.5. Light at twice mains frequency having a form factor of better than 0.96 produces very little stroboscopic effect (5.3.4).

**5.3.3** Terms such as 'flickering' or 'pulsating' do not accurately describe light variations. A light which is *on* for a substantial time relative to the time it is *off* is termed an 'interrupting light', while one which is *off* for a substantial time relative to the time it is *on* is termed an 'occulting light'. Both forms of flashing are used in signal lamps at low frequencies, when the pulses of light are seen individually. When the frequency of flashing of interrupting or occulting light is high enough to exceed the 'visual fusion frequency' of the subject, it is perceived as a steady light. Because of the phenomenon of *persistence of vision*, pulses separated by very short time intervals are seen as a steady light. It is this effect that enables us to perceive cinema and television projections as continuous moving images, though the actual projections consist of a rapid succession of still images and a travelling light-dot respectively.

**5.3.4** Discontinuous light of almost any frequency can produce *stroboscopic effect*, in which a rotating or reciprocating object can appear to be stationary, or moving slowly, or even appear to be rotating in the opposite direction etc. This illusion can cause accidents in the industrial situation. It is a real hazard in the presence of moving machinery; some items (like routing spindles used in making wood mouldings) which have parts rotating very close to or actually at synchronous speed (e.g. at a multiple of the frequency of the supply) can appear to be stationary, and present a serious danger. Apart from counselling the exercise of great care in the use of stroboscopic flash lamps (which are used in industry to measure speeds of rotating and reciprocating devices, or to enable such a device to be examined while moving) no more need be said about them here. But the avoidance of fortuitous stroboscopic effect due to the discontinuous nature of the light output of lamps is a matter of importance (5.3.6).

**5.3.5** Apart from the stroboscopic effect (5.3.6) there are other possible dangers associated with discontinuous light. Some subjects are particularly sensitive to it, and can fall into a light trance or hypnotic state in which they are particularly accident-prone[(12)]. Other subjects who have a predisposition to epilepsy may have a fit triggered off by discontinuous light, particularly at frequencies of around 3 to 8 hz, and sometimes by other periodicities also.

**5.3.6** High intensity discharge (HID) lamps, and fluorescent tubular lamps all have some 'flicker content' in their light output at twice mains frequency (5.3.1), and, exceptionally, (because of a rectifying effect which may occur at one electrode in a faulty lamp) at mains frequency also. Fluorescent tubes and the phosphor-coated HID lamps (MBF, MBFR) have a fairly good form factor because the decay characteristics of the light-pulses from the phosphors tend to be longer than the mains half-cycles, so that the light produced is more or less steady with a ripple superimposed. But HID lamps which give the whole or the greater part of their light output because of the luminance of the arc do not have this 'smoothing effect' on their output, and are therefore more likely to produce stroboscopic effect. In practical applications of lighting, one is not usually concerned with the effect of only one lamp, but of an array or system of lamps; it is then possible to take various practical steps to minimise or remove stroboscopic effect (5.3.7). These steps are important in installations of SON lamps, and of all the metal-halide/mercury-halide series of lamps; in the case of fluorescent tubes, some modern tubes (e.g. the polyphosphor lamps) may have a greater propensity to produce flicker than earlier lamps (Appendix III).

**5.3.7** In any critical situation where stroboscopy might occur, any of the following steps will diminish its effect:

(a) light the moving object with lamps fed from two different out-of-phase a.c. supplies, or from two or three phases of a three-phase supply (2.4.2), or use *lead-lag* luminaires (5.3.8),

(b) select a lamp with low flicker characteristics, e.g. a fluorescent coated HID lamp, or GLS or PAR lamp,

(c) add a local GLS or PAR lamp (preferably fed from a reduced-voltage supply) to augment the general lighting,
(d) (exceptionally) use GLS, PAR or TH lamps fed from a d.c. supply.
**5.3.8** Lead-lag luminaires are twin-tube fluorescent lamp luminaires in which, instead of a normal power factor correction capacitor, one tube is uncorrected for p.f. while the other is over-corrected with a series capacitor. This results in the light pulses from the two tubes being out of phase, the one lagging and the other leading in time sequence. This has the effect of doubling the periodicity of the ripple on the light output amplitude, and considerably reducing the nett ripple, i.e. it improves the form factor of the output (5.3.2) and reduces the propensity of the luminaire to produce stroboscopic effect. It has been noted that on certain designs of lead-lag fluorescent luminaires, the tube on the leading circuit tends to have a shorter life, perhaps due to the electrical conditions it suffers during starting.

## 5.4 Designing lighting to minimise visual error

**5.4.1** Visual errors of various kinds can result in poor quality work and low productivity as well as being a potent cause of accidents. Thus, time and capital expended in producing a lighting installation designed to minimise visual error is likely to be very cost-effective. The following guidelines will aid the designer:
(a) Provide adequate illuminance to enable personnel to move about the workplace in safety, and with a feeling of comfort and wellbeing (2.1.1),
(b) Provide adequate lighting on and around the task to enable the work to be performed efficiently and without strain (1.2),
(c) Determine that the lighting shall be of good quality; consider factors such as an acceptable colour-appearance (4.2), reasonable uniformity, pleasantly graded room luminances (2.2.3), avoidance of disabling glare (1.1.7) and control of discomfort glare (16.3),
(d) Select light-sources that will render colours adequately for the environment (4.2), and select colours for the factory decor that will promote cleanliness and a cheerful workplace (4.1),
(e) Check that the chosen light-sources will be suitable in terms of stroboscopic effect (5.3).

Chapter 6

# Emergency lighting

Many industrial premises were equipped with emergency lighting long before the coming of the Fire Precautions Act in 1971, and the safety value of such lighting is clearly established. Now that compliance with the provisions of the Act is mandatory, it is a management task to determine the nature of the lighting that will best serve the needs of the premises and the organization, and to seek out those methods which will comply with the law, and which are economical, easy to maintain and will give long technical life.

## 6.1 Principles of emergency lighting

**6.1.1** In the UK, all business premises and premises to which the public are admitted are required to comply with the *Fire Precautions Act, 1971*[25]. The Act is concerned both with the prevention of fire, and the provisions to be made for persons to be able to escape from fire or other dangers. The Act's wording contains no direct reference to the subject of emergency lighting, but the direction in Chapter 40 that *'the means of egress shall be capable of use at all material times'* is construed to include that lighting shall be provided along the escape routes to enable persons to use them to reach a designated place of safety. Since the publication of this Act, there has been increasing activity to equip premises with emergency lighting; but there has long been a legal need to make provision for the safety of persons in times of danger.

**6.1.2** Lighting is only one of the many factors which will be examined by the enforcing officer (usually the Fire Prevention Officer of the local Fire Brigade, or an officer appointed by the Local Authority), and only when he is satisfied that the premises will satisfy all the objectives of the Act will he issue the Fire Certificate which authorises the occupier to use and occupy the premises. The enforcing officer has to determine if the escape means are satisfactory, and that the lighting of them is sufficient and suitable, and that it will function during any failure of the mains supply to the normal lighting. This is because the occupier would not be excused his duty to provide lighting because of the failure by any other party to perform any act, e.g. if for any reason the mains electricity supply failed (see Appendix I).

**6.1.3** Escape lighting employs very small illuminances (6.2), and as so few people outside the lighting profession have detailed knowledge about visibility at low light levels, it might be imagined that every enforcing officer would have his own ideas about how the lighting for escape should be engineered, and to some extent this is probably true. But, these days, most enforcing officers will work within the guidelines of *BS 5266 Emergency Lighting* (and its amendment AMD3112)[24]. This is the only authoritative document on the applications aspects of emergency lighting, and has acquired something like the force of law. Under the *Fire Precautions Act,* authorities can take such a document as a code of practice for deciding how the requirements of the Act shall be met. (In the same way, the *CIBS/IES Code for Interior Lighting*[5] is taken as a code of practice in relation to the *Health & Safety at Work Etc Act*[26]; although the Code is not a legislative document, in effect its recommendations are enforceable by Factory Inspectors).

**6.1.4** Emergency lighting is lighting provided to replace temporarily the normal lighting when that normal lighting fails. Its purposes are two-fold:
(a) to enable persons to escape from danger and to get to a designated place of safety; (this is termed *escape lighting*) (6.2),
(b) to enable essential things to be done during a failure of the mains-operated normal lighting; (this is termed *stand-by lighting*) (6.3).

**6.1.5** Emergency lighting luminaires may either consist of lighting units which contain their own secondary-cell batteries and charging circuits—these being termed *single-point luminaires* (6.4.2); or the luminaires may be slave units powered by a *central battery system* (6.4.3). A compromise between these two methods is the *zonal battery system* (6.4.4). Emergency lighting luminaires may be operated in three modes: *maintained, non-maintained,* or *sustained* (6.4.1).

**6.1.6** BS 5266[24] deals with the matter of duration of operation of escape lighting by stating that 1 h duration should be the absolute minimum battery capacity for the smallest premises, and that for larger premises duration of 3 h should be specified, e.g. 1 h duration for buildings up to 10 stories, and 3 h duration for buildings in excess of 10 stories. The BS contains much wise guidance, and many of its recommendations have been proven sound by experience. But, this matter of battery capacity is not one upon which there is universal agreement. Consider the time it takes to search a 10 storey building to ensure that everyone is out; consider what happens when, during that search, an ill or injured person is found on an upper floor; consider the time it will take to descend by the stairs (for the lifts will not be in use), to fetch medical aid, to carry the stretcher to the upper floor, render first aid, and then transport the patient on the stretcher down the stairs to safety, all this is unlikely to be completed in one hour. The specification of battery duration is thus seen to be an important factor in choice of emergency lighting equipment, and it will always be better to err on the generous side and specify 3 h duration, unless there are other provisions (e.g. stand-by lighting from a generator which can be started up (6.3), or mobile lighting (Chapter 14)).

**6.1.7** BS 5266[24] states that routine tests should be carried out to ensure that a system of emergency lighting is in good condition, including operating central battery systems (6.4.3) for at least 1 h monthly, and for single-point luminaires (6.4.2) to be operated from their internal batteries for a short period monthly, with a twice-yearly operation of 1 h. Longer periods of battery discharge may be required at 3-yearly intervals. The BS does not suggest how such partial discharges of emergency lighting battery power can be arranged in continuously occupied buildings, for, if the law is to be followed scrupulously, should the battery be partially discharged and thus unable to deliver the duration specified by the Fire Certificate, then the building should be evacuated until the batteries have been adequately re-charged. One way out of this difficulty could be to provide a larger battery capacity than is needed for compliance with the Fire Certificate, so that a partial discharge for test purposes would still leave the mandatory minimum duration available from the battery. Another method might be to arrange for a contractor to bring a mobile power supply to the building, and to connect it into the circuits to take over the emergency supply duty while the batteries in the installation are being test-discharged and then re-charged. Perhaps this is a service that enterprising emergency lighting contractors will offer. In the case of very large plants, it could be economic to have a mobile power unit (14.2) which would be brought to the buildings in a complex in turn when it was time to carry out the testing of the emergency lighting batteries.

## 6.2 Escape lighting

**6.2.1** The basic requirements for a system of escape lighting are that it shall be available at the instant of failure of the normal lighting or, at most, very few seconds after (6.2.2); it shall provide sufficient light to enable persons to locate the route towards safety and be able to traverse the escape route (6.2.3); and that it shall be in operation long enough to enable all persons to leave the premises (6.1.6).

**6.2.2** BS 5266[(24)] stipulates that, after failure of the normal lighting, the emergency lighting shall be provided within 5 s, though this period may be extended to 15 s if the premises are likely to be occupied for the most part by persons who are familiar to them (this being at the discretion of the enforcing authority). This proviso has little practical value, except for systems where the emergency lighting is provided from generators which take an appreciable number of seconds to come into operation. The idea is a bad one, and it would be far preferable to have a battery which bridged the short time between failure of the normal lighting and the operation of the stand-by lighting. Experience suggests that the seconds immediately following the lighting failure is a time of considerable danger in a factory, and lights which come on within a fraction of second should be provided.

**6.2.3** BS 5266[(24)] lays down the illuminance that shall be provided on escape routes. This is very small indeed, just about the same as full moonlight, and is defined as 0.2 lux minimum along the centre-lines of escape routes, with a maximum diversity of 40:1. This requirement can be met in small rooms by a single luminaire, sometimes the one unit doubling as an illuminated sign as well. Applied to office blocks of cell-offices and corridors, the Standard works very well indeed; but, applied to larger industrial premises which may have many hazards for the escapee, it is doubtful if the BS 5266 recommendations are adequate. Doubts about the adequacy of the 0.2 lux figure stem from experience of the great diminution of light that occurs by the presence of smoke (or dust following an explosion), and the fact that our eyes take an appreciable time to adapt to such a low illuminance following exposure to the illuminances commonly employed in factories (200 to 700 lux general lighting, and possibly much higher values of local lighting). Further, in an emergency, the occupants may be frightened, and their adrenalin reaction will result in their pupils dilating, making them more susceptible to glare and possibly delaying their adjustment to the lower illuminance. Under emergency conditions, the illuminance along the centre-line of an escape route (which may now be cluttered with obstructions) is an unreal concept; what may really be needed is sufficient general lighting wherewith to orient oneself and avoid dangers, plus some luminance ahead towards which one can try to move. The author, having had the experience of being 'blacked-out' while precariously located on a high structure under construction, is convinced that greater illuminances are required, and that these should relate to the risks and to the level of normal illuminance, and therefore somewhat higher levels are recommended (Table 7). The author can also report his experience of being in a burning building, when dense smoke filled the room to within 300 mm of the floor, when seeing, breathing and escaping were possible only by crawling, an experience which makes him have serious doubts about emergency lighting units disposed only at conventional positions high above the floor, from where they can give no guidance nor contribute any light whatsoever to the escape route under dense smoke conditions. It would seem that this problem still requires some practical research and experimentation, perhaps leading to a new kind of emergency luminaire to be mounted close to the floor, or possibly even recessed into it.

**Table 7** Illuminances for emergency lighting

| *Activity* | *Standard design illuminance (lux)* | *Minimum measured illuminance (lux)* |
|---|---|---|
| *Escape along safe and known routes:* Emergency exit lanes, walkways and paths; where the users are familiar with them, or the route is level and not dangerous to traverse | 1.0 or 1 per cent of the normal illuminance, which ever is the greater | 0.2 |
| *Escape along dangerous or unknown routes:* Emergency exit lanes, walkways and paths; where the users are not familiar with them, or the route is uneven or possibly dangerous to traverse, and involves risk of falls, contact with hot or sharp objects etc. | 5.0 or 5 per cent of the normal illuminance, which ever is the greater | 1.0 |

## 6.3 Stand-by lighting

**6.3.1** Stand-by lighting is lighting which will be used during outage of the normal mains-supplied lighting, and its use is not necessarily associated with a critical emergency such as a fire. But, when the mains supply fails, although it may be impossible to continue normal operations, it may be necessary to perform certain essential tasks to prevent possible danger or loss, e.g. to bring dangerous plant to a safe state of shut-down; to hand-wind suspended loads down to take the strain off lifting gear; to damp down boilers or release the pressure in vessels; to stop rotating machinery or apply brakes; to ventilate enclosed spaces etc etc.

**6.3.2** Stand-by lighting is also lighting that can be used for longer periods than those which can be conveniently be covered by storage batteries. Indeed, it is common practice for battery-powered emergency lighting to come into operation immediately on the failure of the normal lighting, but with only sufficient battery capacity to bridge the time until the start-up of generating plant has been achieved automatically or manually.

**6.3.3** It is impractical to provide standby lighting at 100 per cent of that provided for normal working, and the actual level selected may be any value between the minimal levels proposed in Table 5, and the more usual values recommended in the *CIBS/IES Code*(5) and discussed in Appendix II. Where work has been interrupted with the result that there may be significant loss if it is not resumed within a short time (eg the continuous slip-casting or extrusion of materials, the pumping of liquids or slurries that will harden in the pumps or cause flooding, etc) then it is reasonable for the operatives to deal with this exceptional situation by all means available, even if this means working in a lower illuminance than that which would be provided for normal operation. It might be erroneously argued that if the lower level of stand-by lighting suffices for the performance of the work during a mains outage, this lower level would suffice for normal working. The argument is unsound; for while the operatives may react to the situation and perform excellently for a time under the adverse conditions occasioned by the emergency, it is unreasonable to expect that they will be able to maintain output and quality over a longer period. It must be expected that when the illuminance provided is substantially below the recommendations, fatigue will be greater, and its onset swifter (1.2). Further, it must be recognized that simple lack of light is a potent cause of all kinds of industrial accidents (5.1), and that extra precautions should be taken to avoid hazards to personnel. While a low level of stand-by lighting may provide for safe movement, it may be necessary to bring in some form of mobile lighting (14.2) to light the danger spots and the operations where personnel may be at risk. The provision of light-coloured decor (4.1), and marking of vital equipment and movement lanes in white (or self-luminous) paint would be conducive to safety.

**6.3.4** It may be noted that at very low illuminances (say, below 3 lux) the eye's ability to discriminate colour is very limited, and at the illuminances used for emergency escape lighting (Table 5) the ability of colour-vision may be completely absent. Thus, colour-coded signs will not be readable as such, and the use of bold symbols or wording in black on a white background will give better visibility and comprehension under low light conditions.

## 6.4 Emergency lighting luminaires and power supplies

**6.4.1** The classification of emergency lighting luminaires (per BS 5266(24)) as *maintained, non-maintained* and *sustained*, is often misunderstood, as are the relative merits and demerits of each class. *Maintained* luminaires are those in which the lamp operates from the mains supply during normal use, and are automatically switched to the alternative supply on failure of the mains. They may derive their alternative supply either from their own internal batteries (in which case they are known as *single-point luminaires*

(6.4.2)), or from a central battery (when they are known as *slave units* (6.4.3)); zonal batteries may be used instead of a single central battery (6.4.4). Maintained units are lamped with one or more tungsten-filament lamps (possibly krypton-filled) or with one or more miniature fluorescent tubular lamps. This is current practice, but it will be interesting to see if the makers adopt the new 'mini-fluorescent bulb' lamp recently announced; the standard form of this type of lamp operates on 50 Hz, and therefore would not be suitable for use on high-frequency supplies derived from small static inverters.

*Non-maintained* luminaires are generally similar to maintained ones, except that the lamp is not lit when the mains are healthy. Thus, even with the recommended routine testing, the condition of the lamp is not known until the instant it is called on to perform. This uncertainty is greater in the case of tungsten-filament lamps which, of course, fail catastrophically when they fail.

*Sustained* luminaires contain two lamps (or two groups of lamps), one for normal operation, and one for emergency use. Such luminaires have the disadvantages of the other two types described, with the saving feature that the 'normal' lamp could be a lamp of high efficacy, say a small power HID lamp, and thus achieve a current saving.

**6.4.2** Emergency lighting luminaires and illuminated exit signs can be provided with internal secondary-cell batteries to power the emergency lights they contain, together with an automatic charging device to keep the cells in a good state of charge from the mains. Such units are known as *single-point luminaires.* If the emergency light-source is a tungsten-filament lamp, the battery energy may be fed to the lamp as d.c., but if a fluorescent tubular lamp is used, the d.c. battery output will be converted to a.c. by a static inverter. The small inverters used in single-point luminaires are effective for their purpose, but may have power losses of around 20 per cent. Because it is possible to miniaturise the circuit and use a smaller inductive choke for the tube if a higher frequency is used, the static inverter may operate at a frequency of roughly between 18 kHz and 24 kHz according to the particular manufacturer's design, and the output may not be sinusoidal. These units can give rise to radio-interference (10.4) if uncorrected. The printed circuit board (pcb) carrying the static inverter circuit may have the starting device for the tube and its choke mounted on it, and also a charging circuit for the battery. Conversion of the battery output to d.c. enables fluorescent tubes to be used, thus gaining in lumen output and/or duration for a given battery capacity, leading to an overall cost-saving in an installation. Batteries may be sealed nickel-cadmium or sealed lead-acid types, and the pcb may carry circuits to provide trickle-charge, boost-charge, battery charger operation, protection against over-discharge of battery (important for the nickel-cadmium cells), battery capacity indication, and battery state-of-charge indication etc., as well as a test facility. Some pcbs have links which enable the same unit to be converted to work in either the *maintained, non-maintained* or *sustained* modes (6.4.1).

**6.4.3** An alternative to the use of single-point luminaires (6.4.1) is the provision of a *central battery system*. The choice between these two broad methods will usually be made on the basis of first cost, or upon consideration of the forecast maintenance cost for either system in the particular environment. If a central battery system is used, the emergency luminaires are simple *slave units* containing nothing but lamps, and therefore low in cost. A large battery installation must be adequately housed in a location suitable for the distribution cabling, and where the room can be adequately ventilated to minimise fire risk; this can be very difficult or costly for a new installation in an existing building; it is this factor alone that often makes the user choose single-point luminaires. The latter can be easily added anywhere in the premises later without difficulty, whereas if additional slave luminaires are to be fed from a central battery, this may involve long and costly cable runs. The number of slave units that can be added is limited by the capacity of the battery. In industrial situations, the vulnerability of long cable runs to fire or mechanical damage (as well as the need to use large conductors to prevent excessive volts-drop) is another factor causing single-point luminaires to be preferred. However, there is a compromise between the extremes of single-point luminaires or central battery system; this is the *zonal battery system* (6.4.4).

**6.4.4** As an alternative to single-point luminaires (6.4.2) or a central battery system (6.4.3), a zonal battery system could be considered for many industrial applications. In this method, a number of smaller self-contained battery cubicles are used, each providing power for the emergency luminaires in an area. This has several advantages, not the least being that it is often the cheapest method. The luminaires are again 'slaves' containing nothing but lamps, while the batteries are housed in robust, self-contained metal cabinets which can be readily installed in any normally-ventilated room which has a 'normal atmosphere' viz any area that does not have a special fire risk. The problem of volts-drop is minimized, and costly long cable runs are avoided. The reliability of such a system should be as good as any other conventional system, but an extra degree of security can be achieved by connecting a few selected key lights in each area to the battery cubicle in another area (Figure 6.1). This concept can be utilized in the horizontal, or to provide interconnections between floors in multistorey buildings.

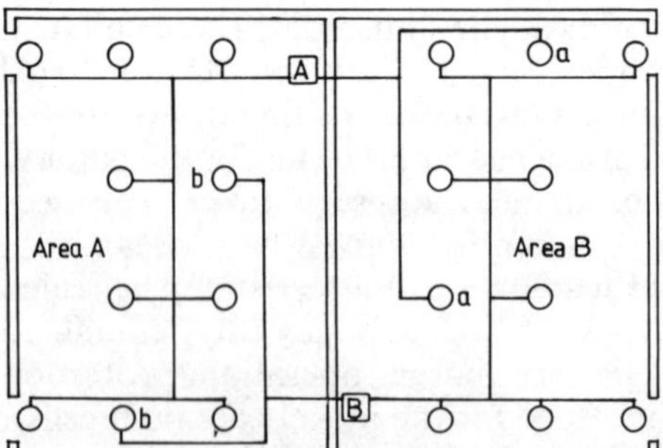


*Figure 6.1* Zonal battery systems. The two areas have individual battery cubicles A and B serving lights in Areas A and B. Additionally, each cubicle supplies a small number of selected luminaires (indicated as a and b) in the adjacent area to give greater security of lighting.

**6.4.5** A new development in the control of stand-by supply systems was discussed in a recent paper[27] and developed at greater length in another book[1]. This is the concept of using a microprocessor to control the fairly complex functions in a system comprising a battery driving a synchronized but unconnected inverter that can be placed in parallel operation with a generator-set also in synchronism. The switching programme (which gives three basic modes: mains, battery or generator supply) must prevent any paralleling of the independent supplies with the mains. Solid-state switches initiated by the microprocessor operate to provide a high degree of reliability, with two stages of fail-safe. The switching is fast enough not to extinguish HID lamps. For large and vital installations, such a system could be economic.

**6.4.6** Another aspect of emergency lighting worthy of attention is its provision, suitably adapted, at outdoor workplaces, and in the provision of lighting on outdoor routes to safety which persons will pass over while escaping from a building. The familiar conventions of BS 5266[24] do not really apply to such situations, but the *Health & Safety at Work Etc Act*[26] does, and this, as well as the *Fire Precautions Act*, might be invoked to compel the provision of suitable emergency lighting for outdoor emergency escape routes.

**6.4.7** To provide emergency lighting in a factory, in offices, or in residential premises, where there is a 'normal dry atmosphere', is really a simple operation, with a good choice of alternative suppliers. But to try and specify suitable emergency lighting equipment for food premises (where the Food Hygiene (General) Regulations and other similar legislation apply) is another matter. There are important applications of emergency lighting which could involve luminaires of hose-proof construction to BS 4533[27], or the need for enclosures suitable for use in special environments (Chapter 10) or hazardous zones (Chapter 11), and selection of equipment and emergency lighting methods for such applications will require much ingenuity on the part of the designer if the results are to be effective, safe, economical and are to comply with all the legal and technical constraints.

**6.4.8** In planning the emergency lighting system, thought should be given to the time which the normal lighting system lamps take to strike, and their hot re-strike time. Tungsten-filament lamps and tungsten-halogen lamps light up and re-light after any interruption almost instantaneously. Fluorescent tubular lamps take a second or two to light up or re-light after an interruption. Mercury-vapour lamps (MBF, MBFR) give virtually no light output on being first switched on, but as the run-up proceeds the light output increases towards the normal, the complete run-up taking about 6 minutes. If switched off and then switched on again, even after an interval of only a few seconds, MBF and MBFR lamps will not immediately re-light, but must first cool down before they re-strike and run up again, a process which may take about 12 minutes. Metal halide lamps (MBI, MBIL, MBIR etc) have similar cold start and hot re-start characteristics to those of mercury-vapour lamps, but particular makes and patterns of lamps can take shorter or rather longer run-up times, a point to be raised with the supplier. High-pressure sodium lamps (SON series) if fitted with internal 'snap starters' may take up to 10 minutes to re-strike when hot, though they may get up to substantially full light output within two or three minutes of starting from cold. SON lamps fitted with external electronic ignitors give a rapid start, coming to substantially full output from cold in a minute or two, and having the capability of giving almost instant hot re-strike. A form of lighting called *bridging lighting* is used to prevent black-out during the starting period of lamps; this may consist of a tungsten-filament lamp running in parallel with the discharge-lamp to give instant light output on switching on. When the light output reaches a predetermined level (or after a fixed time interval) the bridging lamp is automatically extinguished. Such arrangement are really a part of the emergency lighting system, and they provide protection against danger which might occur if lights in a workplace were inadvertently extinguished while the premises were occupied. The illuminance provided by such bridging lighting can be of the order of 10 to 25 per cent of the normal illuminance provided, but, because of the need for instant-starting, the tungsten-filament or tungsten-halogen lamps are used—these having much lower efficacy than the HID lamps forming the normal installation. There are luminaires available which incorporate a bridging TH lamp within the reflector which houses an HID lamp; because the TH lamp is used for only a few minutes at a time, it may last as long as the luminaire without ever needing replacement. A small thermal-delay switch attends to the duty of extinguishing the TH lamp after the HID lamp has run up (Figure 6.2).

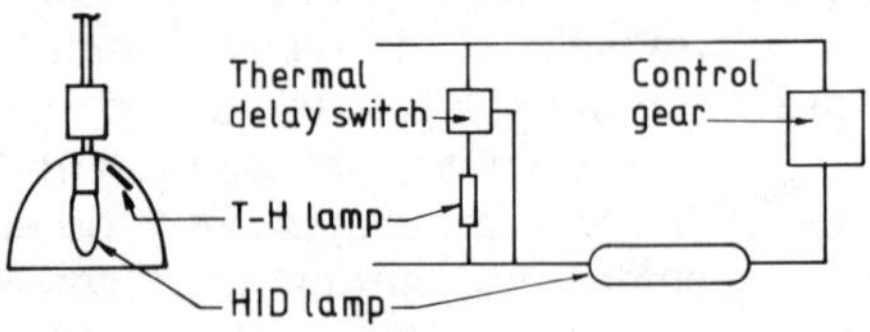


*Figure 6.2* T-H lamp fitted in an HID lamp luminaire to give bridging lighting for safety while the HID lamp is running up. The thermal delay switch extinguishes the T-H lamp after an interval.

**6.4.9** While BS 5266 specifies the general principles of the application of emergency lighting, it does not specify the performance of the equipment. Refer for this to the specifications published by the Industry Committee on Emergency Lighting (ICEL), details of which may be obtained from the Lighting Industry Federation (Appendix III).

Chapter 7

# Procurement of a lighting system

Industrial users of lighting do not usually get enough practice in buying lighting installations to have all the procedures at their finger tips, and study of this chapter may give some useful guide-lines. For the professional specifier, the check-lists may be helpful to ensuring that all important matters are considered. It is vital to appreciate that first there must be a *design brief* in which the objectives and constraints are recognised (7.1); next, an *Outline Lighting Specification* is needed to form the basis for quotations from potential suppliers (7.2); next, the order should be placed on the basis of a *final specification* agreed between buyer and seller (7.3). Finally, after the work is completed, there has to be approval and *acceptance* of the installation by a responsible person before payment for the work is made (7.4).

## 7.1 Setting objectives; recognising constraints

**7.1.1** In the case of a new construction project, it will be part of the duties of the architect and the building services consultant to undertake the specification and design of a suitable lighting installation. If the building is a speculative development, a basic lighting system is sometimes provided, with sufficient capacity in the main wiring and switchgear for the system to be developed or extended to suit the needs of the first occupier. Some new buildings are not provided with lighting in manufacturing areas until they are let or sold, though a basic system of lighting may be provided in office areas and in corridors etc. The nature of the occupancy should determine the kind of lighting system to be employed, and generally the recommendations of the *CIBS/IES Code*[5] will give sufficient guide-lines for initial proposals to be put to the occupier.

**7.1.2** In the case of existing buildings, a very different situation may obtain. Modern lighting installations tend to last longer than the term of employment of typical employees. A manager, on taking up a new job will find a lighting installation in existence, and, in most cases, will simply accept it without analysis or criticism. Indeed, he may expend considerable energy and finance trying to cope with the problems of low productivity, poor quality work, high accident rate, high absenteeism and labour turnover rates, before it may occur to him that at least part of his problems may be attributable to insufficient or unsuitable lighting (Chapter 1). A factory of any size should have a *lightmeter* (Appendix VII) and a copy of the *CIBS/IES Code for*

*Interior Lighting*[5]. It cannot be claimed that a walk round the factory with these two indispensable aids will discover every need for lighting improvement (they will not, for example, tell anything about the glare conditions or the need for improved colour-rendering light-sources) but they

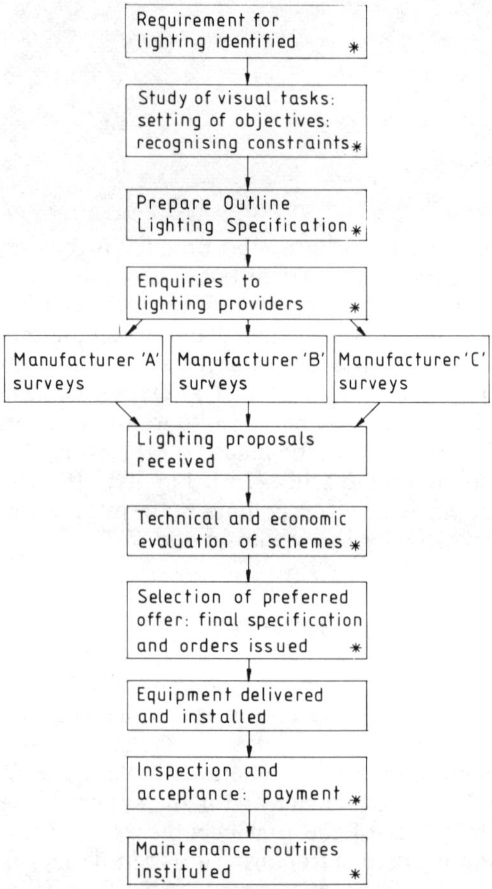


*Figure 7.1* Steps in the procurement of a lighting system. Steps marked * are those in which the professional services of a lighting consultant may be employed.

will certainly deal with the question of sufficiency. Very often, helpful and constructive advice is available from the industrial sales engineer of the Electricity Board, or from the representative of a reputable company of electrical installation contractors, or from the representative or lighting expert of a reputable lighting manufacturer. (Trade Associations, see Appendix VIII; lighting manufacturers, see Appendix X.) Completely unbiassed advice can only be obtained from a qualified person who does not stand to benefit from any business decisions that you make. If the Client does not have a suitably qualified person on his staff to deal with the specification, design and purchase of a lighting installation, recourse may be made to employing a consultant. If going to a general consulting engineering practice, the Client should insist on the work being handled by a person specially qualified in lighting. In the UK, an appropriately qualified person would be a

Member or Fellow of the Chartered Institution of Building Services (CIBS) (Lighting Section). The CIBS maintains a register of such engineers who are of consultant status, ie that are remunerated only by professional fees from their Clients, and who have no other business connections. The help of such an advisor would be of value at practically every stage of dealing with a lighting project for any kind of premises (Figure 7.1) (7.3.4).

**7.1.3** In setting about a project of creating a lighting installation for a new or existing building, the person responsible should define clear objectives as to what is to be achieved, and the criteria by which success with the project is to be judged. Is highest priority to be accorded getting the installation completed in a restricted period of time? (for example, during a factory summer shut-down); or is it more important to complete the task inside a very tight budget for capital expense? Is short-term expenditure on capital cost to be given higher importance than achieving the lowest possible cost-in-use over a review period of three, five or ten years? Will the achievement of the basic lighting parameters (illuminance, glare control, uniformity etc) be regarded as paramount, even if the installation may turn out to be rather unsightly, or if maintenance procedures are going to be difficult or costly (because, for example, certain extensive overhead works of improvement are not to be carried out)? Some objectives that are set will automatically present other objectives which, though worthy, may have to be sacrificed in order to achieve another. Compromises may be possible, but if the objectives are not ranked in order of preference, clean effective decision-making cannot be achieved. An early meeting of the affected parties within the organization is advisable; those consulted will include the budget controller, works manager, production manager, the person responsible for internal installation work and maintenance, etc. First the investigation must find out what is desirable, and secondly what is possible in terms of resources, time, and practicalities. Soon, the investigation will reveal that some desirable courses of action are barred by constraints (7.1.4) and these should be evaluated, quantified and recognized as forming the framework within which decisions about the lighting project must be made.

**7.1.4** Constraints are of two kinds, removeable and rigid. Removable constraints are often those which are imposed by higher authority; e.g., the budget controller has fixed an upper cost for the installation, or the Local Authority has refused planning permission for certain work to be carried out. Such constraints can be removed by logical argument, by appeals, and—above all—by the cogent marshalling and presenting of facts upon which a revised decision may be made. The strongest argument for the removal of a financial limitation is the presentation of a soundly reasoned cost-justification (18.2).

**7.1.5** Rigid constraints are often those which are imposed by the size of the premises, the strength of roof members for weight-carrying, and the absolute limitations imposed by the laws of physics. Some rigid constraints relate to breaches of law which would be caused by a proposed course of action, e.g., to shine a light to seaward 'confusible with a navigation light or likely to be a danger to mariners'. Other legal constraints relate to fire regulations, and instructions of insurers and inspectors of the Health and Safety Executive.

**7.1.6** When the project leader has determined his objectives and set them in priority order, and has recognized his constraints (and determined which are rigid and must be accepted, and those which are possibly removable), it becomes possible to write an *Outline Lighting Specification* (7.2) which will be the framework for the performance of the actual lighting design.

## 7.2 Lighting specification and scheme preparation

**7.2.1** Because every lighting installation is unique, the processes of design and the bases for selection of its equipment cannot be standardised. But, the method of approach will be broadly similar, and should include a review of all the important factors of constraint and objective including the following:
7.2.1a *The structural constraints of the building*
(i) Interior dimensions: length breadth, and height.
(ii) Mounting heights of luminaires in each area, maximum, minimum and optimum.
(iii) Reflectances of interior surfaces and furnishings.
(iv) Obstructions in structure, ceiling or roof features; competition for overhead space from heating and ventilating plant, fire sprinklers, electrical and services distribution runs etc.
(v) Weight of the lighting equipment; load-bearing capacity of the roof structure or ceiling.
(vi) Availability of/desirability of daylight in the interior; combination of electric lighting and daylight (2.3).
(vii) Layout of interior plant and furniture.
7.2.1b *The lighting objectives*
(i) Purpose of the interior; tasks to be lighted.
(ii) Are there very small details to be seen, or visual difficulties because of low reflectances or low contrasts? (Chapter 3).
(iii) Is there need for directional lighting to produce modelling shadows and highlights? Is local lighting under control of the operators needed? (2.2).
(iv) Are there likely to be troublesome reflections? (16.3.5).
(v) What general illuminance is needed? (Appendix II).
(vi) Colour-appearance and colour-rendering requirements (Chapter 4)
7.2.1c *Environmental considerations*
(i) Is the environment a 'normal dry clean environment', or is it hostile to the equipment? Will 'Proof' equipment be needed? (Chapter 11).
(ii) Are there unusually high or low temperatures? (Chapter 10).
(iii) Is the lighting to be integrated physically or thermally with heating/ventilating/air-conditioning equipment? (Chapter 9).
(iv) Need to co-ordinate the lighting with other building services systems? (7.4.1, 12.4.1).
7.2.1d *Lighting design factors*
(i) Luminaire spacing/mounting height ratio.
(ii) Need for enhanced vertical illuminance? (8.3, 17.2).
(iii) Lamp types which meet the colour requirements (4.2).
(iv) Lamp outputs related to powers (efficacies) and suitability for available types of luminaires, mounting heights etc. (Appendix III).
(v) Need to reduce flicker? Danger of stroboscopic effect? (5.3.4).
(vi) Luminous intensity distribution of luminaires; utilization factor (16.2).
(vii) Need for non-standard equipment to suit special requirements?

(viii) Availability (present and future) of replacement lamps and spare parts (13.4).

7.2.1e *Installation and maintenance*

(i) Method of wiring; use a trunking system, or lighting track? (8.2).

(ii) Consider access during installation, and for subsequent maintenance (13.2).

(iii) Maintenance factor to be adopted in calculations; proposed cleaning frequency (16.2).

(iv) Determine method of switching (2.4), and location of switches.

7.2.1f *Safety matters*

(i) Statutory requirements, official recommendations (Appendix I).

(ii) Is there need for emergency lighting, stand-by supplies? (Chapter 6).

(iii) Will colour-rendering suffice to identify dangers, read coloured informative, mandatory and prohibitive signs? (4.2)

(iv) Run-up time of lamps; is 'bridging' lighting needed? (6.4.8)

(v) Safe means of access for all maintenance work? (13.2, 13.3)

7.2.1g *Special design considerations*

(i) Need for intermediate zones of illuminance at entrances? (2.3)

(ii) Special requirements for inspection lighting or viewing luminous tasks? (Chapter 3)

(iii) Need to harmonize methods, equipment or access means with those used in some earlier or later phase of the installation? (13.4)

(iv) Timing of contract; earliest and latest starting and finishing dates; possible hours of work (7.4).

(v) Dangers during installation; any special hazards, e.g. can welding be done in the installation zone? (12.4)

(vi) Need for built-in access equipment? (13.2)

**7.2.2** The scheme preparation will be handicapped from the outset if all the factors listed in 7.2.1 are not considered before any design decisions are made. Preferably, all such data should be briefly recorded in the form of an 'Outline Lighting Specification', and agreed between the Client and the designer. At the very least, the fundamental parameters should be positively identified, thus:

*OUTLINE LIGHTING SPECIFICATION—MINIMUM HEADINGS*

*Part A*

General illuminance (lux).
Plane of measurement for general illuminance (usually taken as floor level or 0.85 m above floor level).
Limiting glare index.
Uniformity of illuminance.
Minimum measured illuminance.
Colour appearance and colour rendering of the sources required.

*Part B*

The lighting provider to be required to state:

The total electrical load including gear losses.
The power factor of the installation.
The utilization factor and maintenance factor employed in his calculations, and the proposed frequency of cleaning and relamping.
That the equipment will be suitable for environmental

conditions at the place of use.
The Ten Year Cost of the installation at agreed present-value rates for energy and labour.

## 7.3 Management of the tender

**7.3.1** The general handling of a tender for interior lighting is not basically different to that for other goods and services except for the twin problems of *two tier quoting* and *skinning.* Two-tier quoting refers to the practice of calling in a lighting provider early in the progress of a scheme, and getting the benefit of his advice e.g. preliminary estimates of cost, drafting the Outline Lighting Specification (7.2.2) and generally treating the supplier as a 'free consultant'. This would be unremarkable if the supplier was then made the nominated supplier and could be certain that the cost of his unpaid efforts for the Client would be covered by his getting the order. Sadly, it often happens that, having got all the problems solved by the helpful supplier, the Client may go out to tender and award the order to another lighting provider. Sometimes the Client may not even include the first supplier on the list of those invited to tender. There is a high probability that another supplier—not having to bear the cost of advising and preparing a scheme—may be able to offer a marginally lower price. The system is unfair to the obliging lighting provider who is contacted first; and it may in time rebound to the disadvantage of the Client, for it is not unknown for suppliers to quote high to such customers. It would be illegal for suppliers to collaborate in this, and in time another supplier comes along who will underbid them. To avoid this unethical and unbusinesslike procedure, it is suggested that in those cases where no consultant is employed, the Client should attempt to devise his own Outline Lighting Specification on the basis of the guide-lines given in this book, and then should call for schemes and quotations against that Specification. Without such a specification, the Client will have difficulty in trying to compare the value of unlike schemes, and would lay himself open to the second malpractice of skinning (7.3.2).

**7.3.2** Skinning refers to the malpractice of lighting suppliers quoting the prospective Client for a lighting scheme, but being deliberately vague as to the parameters of their design. Thus a scheme may appear to be attractive because of its low capital cost, but the Client is not made aware that the overall cost e.g. the *Ten Year Cost* (18.2) will be far higher than another scheme which yielded a better energy consumption or cost of maintenance etc. If selection of schemes was always made solely on capital cost, factories would be lighted again with filament lamps as they were 50 years ago! The data proposed for the potential lighting supplier to provide (7.2.2 Part B) should be matched against the minimum basic data provided by the Client (7.2.2 Part A) plus the Client's other specific requirements.

**7.3.3** While it is expected that the Client will undertake purchasing according to his customary practices and the laws of the country, it would be sound practice to insert an *arbitration clause* into the contract, worded generally as follows or as advised by a lawyer:

*In the event of a dispute between the purchaser and the vendor, either may require the other to enter into arbitration to settle the matter in contention. The procedure shall be that each party shall appoint a single arbitrator to act on its behalf, and the two arbitrators so appointed shall appoint a third arbitrator agreeable to them both who shall act as chairman of the arbitration and shall have an equal vote in any matter voted on by the three arbitrators. The decisions of the three arbitrators as decided by majority vote shall be binding on both the purchaser and the vendor.*

**7.3.4** If the Client decides to appoint a lighting consultant (7.1.2), he may arrange for the work to be carried out in phases according to the needs of the project, thus:

*Phase 1*

Assessment and preliminary report. This may include first estimates of cost, and preparation of an outline specification.

*Phase 2*

Scheme preparation (unless deputed to one or more lighting manufacturers to submit their proposals). In smaller projects, Phases 1 and 2 are sometimes combined.

*Phase 3*

Tender management on behalf of the Client.

*Phase 4*

Assisting with the selection of the successful tender and the placing of orders.

*Phase 5*

Acceptance and commissioning the installation.

In the above work, the consultant will collaborate with and consult those concerned with the project, viz the architect and other consultants, with Health & Safety Executive officers, with Fire Officers and insurers. The consultant's duty is to protect the Client's interests in all matters relating to the project, and to try to achieve the objectives agreed with the Client at lowest cost. Remuneration of the lighting consultant is usually on a *per diem* basis rather than being a percentage of the project cost, but practises vary from place to place across the world. One way of remunerating the consultant that is sometimes used is for him to work on a *per diem* basis until the design brief is settled, and then to quote a fixed fee for the subsequent stages of the project.

**7.3.5** Although there is an obvious advantage in the person who advises on the lighting having no financial connection with any supplier, there are reputable lighting providers who will undertake design work on a fee basis, often with the proviso that if they are the successful tenderers the fee will be rebated against the cost of the installation. With companies of good repute this will work excellently, and provides a method of avoiding two tier quoting and skinning (7.3.2, 7.3.3).

**7.3.6** Quotations often contain alternative offers, or leave items to be decided by the client. It may thus be necessary to provide a *final specification* which may modify some of the data in the outline lighting specification after consideration of the offers received, and this will be the basis for the order on the chosen lighting provider.

## 7.4 Supervision of the contract; acceptance tests

**7.4.1** Because the person responsible for designing a lighting installation is not usually also responsible for supervising the installation of the equipment, the vital need for providing means of safe and easy access to overhead luminaires and wiring may be overlooked until necessity forces attention to the problem (13.1). Commonly, there are unnecessary delays in the progress of the installation work because the activity clashes with some other work going on in the same space; for example, the electricians may be unable to put up scaffolding or other means of access in an area because the floor is being screeded. Or, a convenient system of scaffold platforms erected for the painters (and which would be ideal for dealing with the high-level wiring and installation of the luminaires) is removed at the end of the painting contract, perhaps just a day or two before the electricians arrive. Such wasteful and irritating occurrences might be avoided by obtaining greater collaboration between all parties on the site, and co-ordinating the activities properly. One method of scheduling the timing of activities in the possible sequences is to prepare a critical path analysis diagram (also known as a PERT diagram) (Figure 7.2). This method of forecasting is lucidly explained in Lockyer's book[22].

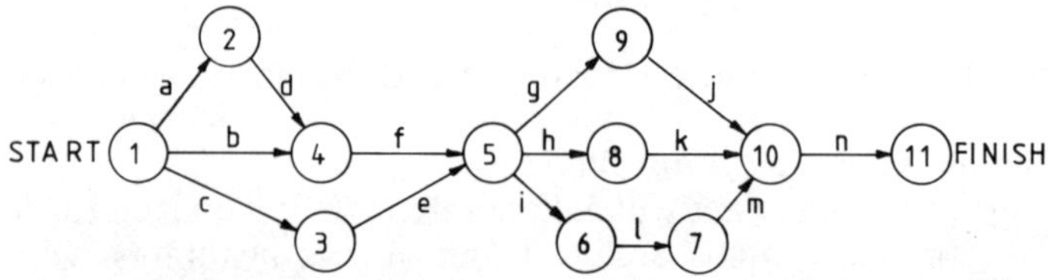


*Figure 7.2* Principles of critical path analysis. The conventions are: the circles are stages of completion; the arrows represent activities; a stage is complete when all the ingoing arrows (activities) are completed; an outgoing arrow (activity) cannot be started until the stage from which it starts is complete. For example, at START (1), activities *a, b* and *c* may be started. Activity *d* cannot start until *a* is complete (Stage 2). Stage 4 will not be reached, and activity *f* cannot commence until activities *b* and *d* are complete. Thus earliest possible and latest permissible completion times are affected by preceding activities. Activities may have 'float', eg, a particular activity may be started between certain dates (head float), and may be completed between certain other dates (tail float). The 'critical path' is the route through the diagram from START to FINISH that takes the longest time (by addition of the durations forecast for the activities represented by the arrows). Thus, any additional time taken by an activity on the critical path will inevitably delay the FINISH, and if extra effort enables an activity on the critical path to be completed in shorter time, this may permit the FINISH to be achieved earlier.

**7.4.2** For the duration of the installation contract there should be provision made for the safe storage of all the materials and components to be used, and for their protection against the weather. Losses by pilfering, and deterioration through damp, are almost entirely preventable by the provision of suitable storage facilities.

**7.4.3** Luminaires designed for 'normal atmospheres' may have very little resistance to rust or corrosion if subjected to damp, this particularly applying to the cheaper ranges of fluorescent tube luminaires. Thus, if such luminaires are stored or installed in rooms which have been recently plastered or treated with water-based emulsion paint, it may be expected that deterioration from damp will result. Indeed, luminaires installed in any unheated industrial building in the winter months are likely to show signs of deterioration very quickly. Either the building should be dried out and have normal heating to keep out the damp, or the installation of any luminaires not adequately rust-proofed should be delayed until just before the area is to be occupied.

**7.4.4** It is a sad reflection on modern life that thefts are such frequent occurrences in all kinds of business. If a newly lighted area is not to be immediately occupied or cannot be made secure against unauthorised entry, it will be advisable to remove costly HID lamps from the luminaires and store them safely elsewhere until the installation is to be brought into use. The high cost of these lamps makes them prone to being stolen, particularly before they show obvious signs of use. Some lamp manufacturers will arrange to permanently mark such lamps with the user's name or identifying symbol.

**7.4.5** It often happens that, in relighting jobs, it is necessary to leave the old installation in operation until the new one is complete. Precautions must be taken to avoid inadvertently energizing incomplete circuits (thus exposing personnel to danger) perhaps by instituting a 'permit-to-work' scheme. No wiring should be energized until it has been checked through with a bell-and-battery or a low-voltage continuity tester, and the insulation between lines and from lines to earth has been checked with an insulation test instrument as laid down in the Wiring Regulations[15] and the appropriate certificate completed. Luminaires which contain ignitors or other solid-state circuitry (as control-gear components or as part of emergency lighting circuits) will be damaged by over-voltage or reverse-voltage applied during insulation testing, so such circuits should be shorted-out before an insulation test is performed.

**7.4.6** At the completion of the electrical installation work, the name and address and telephone number of the electrical installation contractor and of the local electricity board should be permanently recorded and displayed near the main switch of the installation. It may be a requirement of the local Fire Prevention Officer who carries out inspection of the installation in connection with the granting of the Fire Certificate (*Fire Precautions Act,* see Appendix I) that a diagram of the circuits and the location of the switches be displayed also, or that a copy be readily available to the Fire Brigade in an emergency so that circuits can be rapidly isolated by the Brigade.

**7.4.7** If the lighting provider who supplied the luminaires has provided the required data, it will be possible to carry out a lightmeter check to see if the installation produces the illuminance required by the Specification. This is a contentious matter, over which purchasers are often short-changed by unscrupulous providers (7.2.2). If the initial lumens, the lighting design lumens and the end of life (terminal) lumens per lamp are known, as well as the utilization factor used in the design (Chapter 16), the light loss factor can be determined to enable a forecast to be made of the future minimum illuminance or service illuminance by the method given in the *CIBS/IES Code*[5]. In this way, it is possible to compare the initial illuminance (measured on the day the installation is commissioned) with a theoretical value, and forecast what will be the service illuminance etc in the future. One must overcome the misleading effect due to the 'bonus lumens' when the lamps are new by delaying measurement of the illuminance until the lamps have operated for 100 h (16.2.5). Sometimes, a 10 per cent or 15 per cent 'hold-back' of the payment to the lighting provider may be a feature of the purchase contract, the final payment to be made when the installation has satisfactorily passed these acceptance tests.

Chapter 8

# Installation design: practical considerations

In this chapter, the reader is led towards the ideas which will link theoretical concepts of lighting to the practicalities of applying those concepts to the premises. The general ideas of how to achieve good lighting thus become specific to an actual application, and must be adapted and harmonized with the work of the factory and its physical environment. Much of the material here will be of value in discussions with the lighting providers and installation contractors for industrial lighting schemes.

## 8.1 Choice of lamps and luminaire types

**8.1.1** It is an objective of this book that the user will be guided towards logical selection of his lighting equipment. Long experience indicates that it is common for industrial managements to simply repeat a method of lighting which exists elsewhere in their works when an extension or refurbishment of lighting is needed; worse still, some managers, having seen a successful lighting scheme in another works, will instruct that a similar pattern be applied to their own premises without any analysis of visual function and specific requirements. It is not unknown for managers responsible for the purchase of a new lighting installation to pick a luminaire or lamp type from a manufacturer's catalogue by fancy, or to accept unquestioningly the first suggestion of a representative of a lighting company. Perhaps this irrational method of selecting the lighting method is prompted by ignorance, the manager feeling that the subject is too simple to bother with—or, conversely (because he has read but not understood one or two technical articles)—that the matter is too complex and is better left to others to decide[4]. Lighting is not as simple as it looks, and in most cases some expert advice would be useful. But, where shall the user go to for that advice? If he seeks advice from any lighting manufacturer, it is most likely that the answers he is given will be influenced by the goods that the manufacture has to sell. And, the user may not feel that such a routine matter as buying some lighting justifies paying the fee for a consultant (7.3.4). It should be remembered that the price paid for a luminaire is about one quarter of the total cost of that luminaire over 10 years, and perhaps only 6 to 10 per cent of the total inflation-adjusted cost over that period including lamp replacements, maintenance and energy cost, and a notional allowance for interest in the capital value depreciated over 10 years (18.3). Lighting decisions are important in that they commit the organization to future expenditure.

**8.1.2** It has been shown that the quality and quantity of lighting provided in factories is an important factor in the environments of workers, affecting their attitudes to their employment (1.1); it is clearly established that good lighting can be a significant benefit to productivity (1.2), and that it is important to industrial safety (1.3). Overall, there should be a calculable cost-benefit from a well chosen and well maintained lighting scheme (1.4). Further, although the cost of energy for lighting in a typical factory represents only about 6 per cent of the whole energy bill, wasteful practices must be avoided, and the effect of the efficacy of the lamps on the heat-balance of the building should be considered (9.1). This clearly is not an area for making minor economies, for such savings as may be made by skimping on the lighting are likely to be cancelled out and turned into losses (18.1). While efficacy of lamps is a very important factor in determining what the total cost of that installation will be over say, a 10-year period, it is not the only reason for selecting a particular type of lamp. The colour properties and other factors of design application may also be of importance (4.2).

**8.1.3** If a simple guide-line for selecting a type of lamp must be stated, it could be: *'Select the highest efficacy lamp which will satisfy the colour-rendering and colour-appearance requirements of the application, and employ the highest practicable power of that lamp.'* Similarly, a simple guide-line for selecting the luminaire could be: *'Select the type of luminaire suited to the preferred lamp type that will satisfy the light-distribution requirements of the application and which will give long and safe service in the environment of operation with minimum maintenance cost.'* The methods for translating these important decisions into management action are given in Chapter 7. The reference to 'environment' in the guide-line is of great importance; for many industrial workplaces have environments which are hostile to lighting equipment not suited to them, leading to low efficiency, frequent breakdowns, high cost of maintenance and possible danger (Chapter 11).

## 8.2 Suspension and wiring systems

**8.2.1** Choice of suspension and wiring method should relate to the building structural details, to the mounting height, the type of luminaires and the maintenance methods proposed. Where there are exposed joists, the obvious step of suspending the luminaires from them is not necessarily the best; it may happen that the spacing of the joists nearly coincides with the required spacing of the luminaires , but joist-mounting usually means that the end luminaires are too far from or possibly too close to the end walls. It is usually possible to arrange the pattern of HID lamp luminaires so that the luminaires do not coincide with the joists, which offers the possibility of mounting the luminaires so their bottom rims are level with or slightly above the joist level, thus giving some protection to the luminaires in areas where fork-lift or clasp trucks are employed.

**8.2.2** The traditional method of wiring in steel conduit may be more costly than the use of *lighting trunking*. The trunking provides a rigid support for the luminaires, and the conductors within the trunking provide the means of connection. There are systems suitable for carrying HID lamp luminaires, and others in which the control-gear for fluorescent lamps is housed within the trunking. Trunking facilitates access, in that if it is suitably fixed and braced, ladders may be leaned against it (but preferably fitted with hooks to prevent slipping) (13.2). Combined trunking may provide distribution for power and lighting.

**8.2.3** Another possible method of mounting and wiring is the use of *lighting track*. The track supports the luminaires, and has two or more electrical conductors to which the luminaires are connected by a fitment. Track of suitable design can carry HID lamp luminaires, and is especially suitable for overhead-mounted localized or local lighting (2.2). In multi-circuit track, switching of the individual circuits presents no problems.

**8.2.4** Luminaires may be detachable for maintenance (13.1), and winding-gear may be used to lower them to floor level (13.3). The use of captive fastenings, and safety-chains to prevent items falling, is recommended, and is mandatory in places where foodstuffs are stored or handled (19.1).

**8.2.5** Rigid mounting of suspended luminaires is only suitable for short drops (say around 300 mm), longer drops being fitted either with conduit swivel-plates, paired hooks on conduit, or the suspension may be by chains. The contractor should assure himself that cables required to flex shall be of a standard that will not fail in service, and cables may be housed in metallic flexible hoses for protection against the environment, mechanical damage and possible attack by rodents. If chain suspension is used, the ordinary grade of bent-wire chain-link will not satisfy industrial use, and the chain should be heavy-gauge welded-link pattern.

**8.2.6** Mineral-insulated metal-sheathed cable is extensively used in factory installations, and has much to recommend it. It is cheaper per metre installed than pvc cables in any form of metallic conduit, and the installation labour cost is low. However, it must be remembered that the copper or aluminium sheathing is not as physically robust as steel conduit, and exposed runs should be suitably protected. The entry of moisture into this type of cable reduces its insulation value and will lead to breakdown; therefore cut ends of mineral-insulated metal-sheathed cables should be sealed at once if they are not to be immediately connected using the proper fitting. (Bending the end over and hammering it down flat is fairly effective for a short time in a dry location.)

**8.2.7** Consideration should be given to the effect of vibration due to oscillation of steel-framed lightly-clad buildings, and the effect of gantry movement. Vibration can seriously affect lamp-life (11.4).

## 8.3 High rooms; rooms with gantries

**8.3.1** In rooms having a luminaire mounting-height substantially greater than 6 m, it may be found that with the use of 'high-bay' or concentrating luminaires, e.g. of low BZ Number (16.1.4)—it is likely that the horizontal illuminance $E_h$ will greatly exceed the vertical illuminance $E_v$, with the result that the vertical faces of objects will tend to be underlit. If luminaires with wider, less-concentrated beams are used, the wastage of light on the walls will make the scheme uneconomic (Figure 8.1(a)). The use of *cross-lighting* to enhance the vertical illuminance has much to recommend it (Figure 8.1(b)).

**8.3.2** A common problem is in relation to overhead gantry cranes, which can obscure a significant amount of light in high-bay HID installations. In some cases the user mounts additional high-bay units on the underside of the gantry to compensate for this, but this has the serious disadvantage of

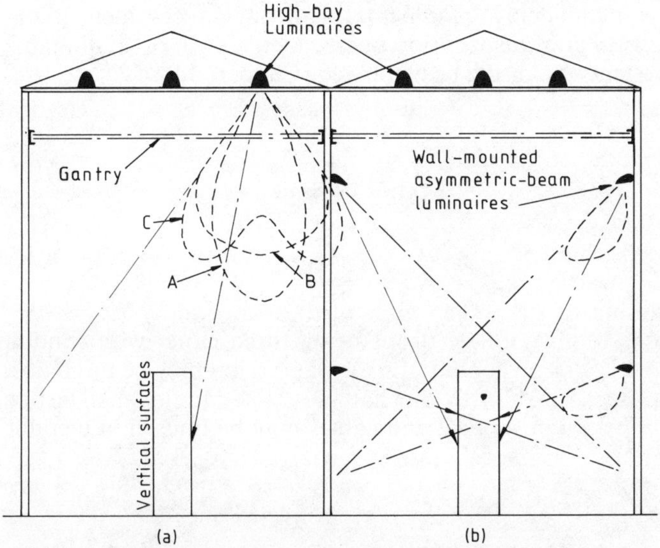


*Figure 8.1* Enhancing vertical illuminance in high rooms. (a) Conventional high bay concentrating luminaires achieve the required $E_h$ but fail to light the vertical surfaces of objects in the area satisfactorily. Typical distribution curves are shown, A — high bay luminaires, B — wider angle (dispersive) luminaires, C — Batswing or 'trouser-leg' distribution. (b) The addition of wall-mounted cross-lighting luminaires (mounted below gantry height) will augment the $E_v$.

producing moving shadows at the working plane, and this is considered to be a possible cause of accidents (5.1). While the use of a suitable spotlight on the hook may have some advantage (of a kind which can be switched on and off readily, e.g. a tungsten-filament or TH lamp, so that the lamp is not lighted when the crane is moving) it is preferable to arrange that the lighting in areas where gantries are used produces satisfactory visibility without such additional lights. The problem of obstruction of the high-bay luminaires can be minimized by mounting the luminaires in staggered formations or inclined transverse rows so that only one or two overhead luminaires are obstructed at any crane position. With suitable precautions, the gantry may be used to gain access to the high-bay luminaires for maintenance purposes (13.2.11).

## 8.4 Temporary interior lighting systems

**8.4.1** Temporary lighting installations, often improvised hastily to cope with some unexpected difficulty, are a potent cause of accidents such as electrocution and starting fires. The word 'temporary' must never mean 'of a lower standard of safety'. All installations must be tested before being put into service, and shall comply with the 'Wiring Regulations'[15] as regards insulation resistance, correctness of polarity and earth-continuity etc.

**8.4.2** The possible need for temporary lighting for special occasions, e.g. for installation of new plant, for descaling boilers or major periodic maintenance operations, should be anticipated, and suitable lighting equipment purchased and held in readiness. It is strongly recommended that a *reduced-voltage system* of distribution be installed (14.4), or, failing this, some other safe system of lighting be employed (Chapter 14).

Chapter 9

# Thermal, ventilation and energy considerations

The practices of 'wise use of energy' are not only necessary because industry must respond to urgent needs arising from world-wide concern about future availability of energy; they represent very good economics for industrial concerns, even in the short term. The principle areas in which energy conservation in buildings may be carried out are readily identified (9.1), and the factor of lamp efficacy is clearly an important area of cost-saving (9.2). In many factories, improvements in lighting associated with the installation of false-ceilings in high-roofed buildings will bring valuable environmental improvements as well as energy savings (9.3). There is no difficulty in getting expert advice on building energy matters, and many techniques which were developed for commercial buildings may be applied to industrial premises also (9.4).

## 9.1 Energy conservation in buildings

**9.1.1** Buildings can be comfortable and acceptable to their occupants, even though the window area is small, or even if the buildings have no windows, provided that the electric lighting is devised to produce good visual conditions and a humane environment. Small windows, or no windows, result in there being less loss of heat during cold weather, so that heating bills are reduced. In the summer, heat gain is smaller, so fan-ventilation or air-conditioning can produce very comfortable conditions. In buildings with limited fenestration plus air-conditioning, better control can be exercised over cleanliness, and there can be better limitation of noise entering the building. Similarly, the relative humidity within the building can be controlled to produce good conditions of personal comfort[32]. Some of the ideas which have been successfully used to improve thermal conditions in commercial buildings can now be applied to industrial buildings, often in conjunction with modernization work in older high-roofed buildings (9.3). The performance of a detailed energy audit for the premises will often reveal that significant savings in energy may be made in return for relatively small capital outlay. The practice of 'wise use of energy' is thus not only seen to comply with urgent needs pressed upon us by the world-wide concern about availability of energy, but it is good economics even in the short term.

**9.1.2** It has to be admitted that most of the application of scientific energy-conservation ideas to buildings have been to commercial premises; and it also must be admitted that most of the few applications of these ideas in industrial buildings have been under sponsorship of The Electricity Council, e.g. the Wallsend Meter Testing Station,[28] or under Government sponsorship to produce case-histories. It would be wrong to believe that such ideas as windowless factories, heat-recovery systems, air-handling luminaires, heat-pumps etc, are only for organizations that are willing to spend money purely for prestige purposes and who feel that it is good for their image to be involved in the creation of sophisticated building projects. It is probably true that industrial managements are far more interested in increasing profits by money-saving schemes than in getting involved in more complex designs to save a little energy. Energy-saving projects are seen more as political than economic, and many managements simply will not devote time to studies to see if some of the new ideas on energy could in fact benefit them. However, it seems that this situation is beginning to change; there are examples of considerable capital expenditure by industrial organizations on schemes of energy conservation, in most cases allied to improvements in the working environment calculated to improve productivity and staff relations. At the time of writing, Bank Rate is 13 per cent, and the rate of annual inflation is 15 per cent; against this background, every capital investment must be weighed against the alternative of earnings in the Deposit Account that are forfeited, plus the fact that the cost of owning an installation is probably going to be three to four times the straight-line forecast without allowing for inflation over ten years[33]. But energy-conservation steps in relation to lighting and building energy do make substantial savings; and it can only be a counsel of prudence to advise that the economic possibilities are thoroughly investigated whenever there is need to update or refurbish any industrial building, and, particularly, in the case of new industrial building projects, to require that the architect, building services engineer and others concerned with the project do not miss the opportunity to investigate and weigh up alternative methods of lighting and heating that may be cost-effective as well as energy-saving. A 'low energy' building will probably be more costly initially, but over its life it is likely to achieve significant reductions in running costs compared with conventional energy-hungry buildings.

**9.1.3** With good reason, there is much public and governmental concern about energy; the world situation demands that we do not waste it. Savings in energy are in line with public policy as well as being of economic value to the saver; but it will be as well to record here some advice about the energy-economy aspects of lighting, and at the same time to refute some widely-held and mistaken ideas as to what constitutes real energy saving in lighting:

*(a) Can a significant energy saving be made by switching off lights for short periods when not actually required?*

It is unlikely that any nett saving will result from switching off lights for periods of less than 10 or 15 minutes, but it is clearly economic to switch off for periods of about an hour or more (2.4.3). Switching out might expose personnel to danger; in the case of HID lamps (other than those fitted with electronic ignitors) there will be some delay at switching on again before the lighting is available (2.4.4).

*(b) Is it economic and safe to operate twin-tube fluorescent lamp luminaires on one lamp only?*

There have been reports of luminaires overheating from this cause, but in most cases it is probably safe. It is more economic to change the luminaires to

a type or spacing that will give the required illuminance in the area concerned.
*(c) Is the cost of reflectors or refractor-enclosures justified, or are not bare-tube battens more economic?*
Although the Light Output Ratio of bare-tube battens is high, they tend to distribute a large proportion of their output in directions that do not contribute to direct lighting of working plane. Slotted-top metal reflector luminaires are about 25 per cent more efficient at putting the light on to the work. Further, unshielded lamps cause direct glare, and thus reduce the 'seeing value' of the available light. Prismatic refractor-enclosures direct the light usefully, and also reduce the brightness of the luminaire in directions of view.
*(d) In view of the need for economy and wise use of energy, are the recommendations of the CIBS/IES Code[5] justified? After all, people did use to manage with far less light.*
In industrial workplaces, the aims of economy and wise use of energy are more likely to be achieved if the work is not hampered by poor lighting (1.4). When a piece-part is rejected as faulty, what is thrown in the scrap-bin is not only the materials and labour used to the point of rejection, but also all the energy that was employed in the manufacture of the part. Of course, once upon a time, people managed to perform simple trades by the light of candles and oil-lamps; but they were not carrying out the technical tasks of the present day, nor stressed by the pace of modern industry. Lighting is getting progressively cheaper in monetary and real terms: modern lamps last longer, and give more lumens per unit of energy consumed; and modern luminaires are cheaper, and make more efficient use of the light. The cost of energy has not risen as fast as most other industrial costs. The result is that every industrial concern can afford to have lighting to the standards of the *1977 CIBS/IES Code.* The levels recommended in the *Code* merely represent good present-day standards, based on scientific research, much field experience and taking account of the preferences of users; and thousands of users have found that the standards of the *Code* aid their industrial and commercial operations and contribute to the profitability of their organizations, as well as to the safety and wellbeing of their workers. Thus, well designed and efficient lighting is seen to be a wise use of energy.
**9.1.4** Currently, the UK Department of Energy operates an 'Energy Saving Scheme' whereby occupiers of industrial premises can have an *energy audit* carried out by a listed consultant, with a valuable Government subsidy on the consultant's charges.

## 9.2 The importance of lamp efficacy

**9.2.1** The luminous efficacy of lamps (lumens per watt, lm/W) is really an index of value for money. The higher initial cost of HID lamps, and the higher cost of lamp replacements, compared with, say, tungsten-filament lamps, is justified by the reduced overall cost of lighting over a period of time. While the proportion of the energy usage in an industrial building attributable to lighting is small, in thermally efficient buildings the heat from the lighting can contribute significantly to the heating load required (9.4); but lamps of lower efficacy add to the summer cooling load, as well as being more costly to operate.

**9.2.2** In practical lighting installations, there are inevitable losses of lamp lumens between the lamp and the working plane (16.1). To get an appreciation of the magnitude of these, consider that in a typical industrial interior lighted with common types of industrial luminaires, the Utilization Factor might be around 0.66, and the Maintenance Factor around 0.75. This results in a lamp lumen requirement of

$$\frac{1}{\mathrm{UF} \times \mathrm{MF}} = \frac{1}{0.66 \times 0.75} = 2.02$$

In other words, for every lumen per square metre (lux) on the working plane, about twice that number of lamp lumens must be provided. While good luminaire design and good maintenance will limit this proportion of non-effective light, the only way that the economics can be substantially improved is by using lamps of higher luminous efficacy.
**9.2.3** As an illustration of how lamp efficacy affects the cost of lighting, consider that in the period 1950 to 1980, energy costs have typically risen in the UK from around 1 Old Penny (0.42p) per kWh to around 4p/kWh, a tenfold increase. Yet, because lamp efficacies have increased substantially over the same period, one now gets about four times as many lumen-hours per £ expended than 30-years ago. The process has been aided by the extension of lamp lives, and the cost of luminaires has been restrained by improved optical and thermal designs coupled with the benefit of modern manufacturing methods. It is, however, the improvement in lamp efficacies that has made lighting continue to be one of the few commodities that have reduced in real cost over the period; and, although wages continue to inflate, the cost of lighting forms a progressively smaller fraction of factory costs. In choosing lighting equipment, the dominant factor in the cost of owning the installation over a review period of, say, 10-years, will be lamp efficacy (8.1).

## 9.3 Lighting and ceiling structures

**9.3.1** Surveys have shown that only about one-third of factories are housed in purpose-constructed buildings; most factories have adopted and adapted existing buildings to suit their purposes, or have purchased standard prefabricated buildings. The result is that rather more than three-quarters of all factories have greater roof height than is actually needed by the work performed. In quest of improving the working environment and making economies in energy consumption, many factories have installed false-ceilings under their apex-roofs or northlight-roofs, reducing the ceiling level to eaves height or lower. Where the floor space is broken into smaller areas by partitions, the lower ceiling makes these smaller spaces far more comfortable, and reduces the travel of noise from one area to another. The purpose for installing a false-ceiling may be to reduce the air-volume, and thereby economize on the heating cost, and in the cost of forced-draft ventilation or air-conditioning. In food factories, false-ceilings may be installed to improve hygiene; while the effect of noise-reduction on fitting an acoustically-dampening false ceiling in a noisy works can be dramatic.
**9.3.2** Where a factory roof is fitted with laylights or northlights which

admit daylight, many of these windows will admit some sunlight too, making the building over-heated and very glareful in the summertime. Some so-called northlight-roofs are incorrectly orientated, with the result that shafts of sunlight penetrate at one or other end of the day. The writer has often visited factories where rooflights are 'green-washed' every spring, or where canvas blinds are suspended under the rooflights to reduce this problem, and, almost invariably, the general lighting has had to be on during all daylight hours, so the possible savings due to the use of daylight have been entirely negated.

**9.3.3** In about three-quarters of factories visited, it was found that the electric lighting was switched on, irrespective of the amount of daylight coming into the interior. Often, the use of the electric lighting was necessary to offset sky-glare from side windows, and to provide a greater horizontal illuminance. When the layout of the building is freed from the constraint of relating to daylighting from side windows, there can be more efficient usage of space. The value of the light entering the space from roof windows (i.e. the cost of replacing it with electric light) is minor, and is not a factor that should unduly affect decisions about factory space utilization. On the other hand, the heat loss and unrequired heat gain through such windows can be significant.

**9.3.4** There are established methods of augmenting daylight with electric lighting (2.3), but it has to be admitted that such systems are often left with 'all lights on', irrespective of the daylight available. Daylight is variable, and workers do not want to have to switch lights on and off frequently according to passing cloud formations, though photocell-controlled automatic switching can be employed (2.4.6). Rooflights do not have the 'social value' of wall windows (you cannot see out of them, and the general use of wired-glass prevents even seeing how the weather is), and can be dispensed with in most cases without inflicting psychological traumas on the occupants. It is, however, absolutely necessary to replace the rooflighting system with a first class system of electric lighting. In some instances, a translucent false-ceiling is fitted, so that some of the available daylight filters through. It is possible to position a proportion of the general lighting luminaires above such a false-ceiling, and, although the utilization factor of such luminaires is not high, this can produce a good visual effect. It must be commented that it is often very difficult to clean the upper surface of suspended translucent false-ceilings, and in typical factories they soon become more or less completely occluded with dirt and dust. For this reason, fire-resisting open louvre grid ceilings may be more suitable in industrial areas.

**9.3.5** Turning now to practical considerations, advice should be sought from the *Fire Precautions Act* enforcing officer before making final decisions about fitting false ceilings in existing buildings. He may have stipulations to make about venting of smoke in case of fire, and may impose specific requirements regarding the fire-retarding properties of such a ceiling, according to the work being performed in the area and its fire-risk. If sprinklers are to be fitted at false-ceiling level, additional sprinklers may have to be fitted at higher level too. There are systems of electrical and lighting trunking which are adapted to support a false-ceiling between the trunking runs. Thought should be given to how the lighting installation will be maintained; in some cases it is possible to gain access to the luminaires through the ceiling void, with walkways being provided for this purpose.

**9.3.6** Lowering the nett ceiling height by adding a false ceiling may act as a constraint on the power of HID lamps that can be employed at the new ceiling height. Until quite recently, it was a fairly rigid rule of thumb that HID lamps would not be used at mounting heights below about 6m[4], but, with the development of new kinds of luminaires, much lower mounting heights for HID lamps are now commonly employed, down to as low as around 3 m in some cases. The design of such installations can make some difficulties for the lighting engineer, but, by the use of luminaires giving better optical control, nowadays it is feasible to employ HID lamps at lower mounting heights than were thought practicable a few years ago.

## 9.4 Integrated environmental design; heat balance

**9.4.1** A building that is thermally efficient, i.e. that is of high mass, or is well insulated, will retain much of its internal heat gains from processes, lighting energy and the heat given off by occupants. Thus the building may come into heat-balance at a lower temperature. This means that in cold weather, less make-up heat has to be added to provide environmental comfort. In a thermally efficient building, this would be achieved with heat-gains of 43 to 54 watts per square metre (4 to 5 watts per square foot) of building area. When it is realized that 32 $W/m^2$ (3 $W/ft^2$) is obtainable from 500 lux of (fluorescent) lighting alone, it can be seen that it would not be difficult to obtain heat balance at an outside temperature in the range of 4 °C to 8 °C, producing a building that virtually heats itself[28]. It is an apparently paradoxical situation, in that the objective of good lighting design is to try and achieve the required lighting with minimum energy usage; yet in thermally efficient buildings, it would seem that efficacy of lamps would not be so important as it is in other buildings. This is not true, and the picture is changed by considering the thermal gains and losses of the building throughout the year. In the summer, excessive heat gains may have to be dissipated by ventilation and cooling, with extra usage of energy. One of the first applications of the concept of 'integrated environmental design' (IED) to an industrial building was at the Meter Testing Station at Wallsend, owned by the North Eastern Electricity Board, and it has been an outstanding and most interesting project, the principles of which could be applied with great value and economy in many types of industrial buildings. For the full IED concept to be applied, it is necessary that the decisions be formulated before the architect draws single line—the essence is the integration of ideas on lighting, heating, cooling, ventilating etc from the outset[28].

**9.4.2** Heat exchangers can be used to recover the heat in extract air in air-conditioning systems. This extracted heat can be used to pre-heat the make-up air, and so raise the thermal efficiency of the system[30]. In office installations, extraction of air through the luminaires is sometimes practised, the effluent air collecting a large proportion of the heat emitted by the lamps as it passes through, with the result that fluorescent tubes being so cooled will emit extra lumens—perhaps as much as 10 per cent increase in light output may result.

Used in conjunction with a heat-pump to recover the waste heat, savings of up to 40 per cent of the energy cost for heating and cooling a building can result[30, 31]. So far, these ideas have been practised mainly in office buildings and shops; but, for clean industrial areas where there is a significant heat gain from process heat and the heat of occupancy, the concepts should certainly be explored, both for new buildings and for existing ones. In the UK, advice is available free from the area electricity boards, and from the Air Conditioning Advisory Bureau of The Electricity Council whose computer programme on building energy (BEEP) can usually identify those cases in which heat-recovery is likely to be practicable and economic (see also 9.14).

Chapter 10

# Lighting for special industrial environments

This chapter gives some examples of special requirements for lighting equipment to comply with the environmental conditions and constraints. The resources of the lighting industry in the UK are adequate to cope with even the most unusual demands for performance of lighting for all kinds of environments, but it is important that such applications should not be subject to improvization and compromise, but that the constraints on the performance of the lighting system are recognized early, so that expert advice can be obtained (7.1, 7.2). Lighting can be devised to operate in extremes of temperature (10.1, 10.2), in environments of clinical cleanliness (10.3) and without causing radiointerference which could affect sensitive equipment as well as radio/TV reception (10.4).

## 10.1 Lighting in high ambient temperatures

**10.1.1** In factories operating hot processes, it is common to find that the ambient temperature at luminaire level is considerably above the normal range in which standard lamps, control gears and wiring are designed to operate. Suitable lighting equipment is usually available, though at extra cost, to enable lights to operate at all temperatures in which a man can work, say up to 35 °C (95 °F), but this statement should not be taken to mean that standard lighting products will be satisfactory at such temperatures. The optimum wall-temperature for standard fluorescent tubular lamps is 38 °C (100 °F), and if the tube runs hotter or cooler than this the light output will be reduced. The wall temperature of HID lamps is around 120 °C (250 °F), so the lamps are not much affected by high ambient temperatures if operated in suitable luminaires. But, at such elevated temperatures, the control-gear will tend to give short life, capacitors in particular may fail early (and some types of capacitors will leak oil, with a possible fire-risk), and luminaire connections must be such that the insulant of cables is not damaged by heat. (Typical modern cable insulants for normal temperatures will show signs of softening or creepage at 95 °C, and can start to flow at around 105 °C. Thus specification of temperature at which cables will operate is an important item for safety.)

**10.1.2** In rooms with flat ceilings or well insulated roofs, it is possible for static air above the highest window opening to be very considerably hotter than the air in the normal occupancy zone, say between floor level and up to 2 m above the floor. In fact, such pockets of hot air may be unsuspected until low light output from fluorescent tubes, or early failures of control-gear draws attention to them. Similarly, luminaires located in a ceiling void above a false-ceiling (9.3) may be subject to overheating if the void is not ventilated. In some factories operating hot processes, a column of very hot air may rise, which while not greatly affecting the average temperature of the space, may seriously overheat luminaires upon which it impinges. The author has visited factories where the temperature in the roof space above hot plant was so great that it was impossible for staff to enter it to attend to normal relamping and maintenance of luminaires, such operations having to wait until the plant was shut down.

**10.1.3** Most complaints about the failure of lamps, luminaires and control gear to perform well at elevated temperatures are due to the purchaser not informing the supplier of the environmental requirement. To special order, lighting providers can supply chokes and capacitors rated for higher than usual temperatures, and they will be able to advise the user (on the basis of tests made to BS 4533[27]) whether there will be any special requirements as regards the connecting cables. In some cases the manufacturer may stipulate that a special heat-resistant grade of cable is employed. Incidentally, in many HID lamp luminaires, such heat-resistant cable is employed for the internal wiring; this should never be substituted with any other grade of cable.

**10.1.4** Although a fluorescent tubular lamp may be operating in an elevated temperature, it can be made to give a satisfactory lighting output if its internal gas-pressure can be reduced. Even applying a small area of cooling will be successful in this, and various ingenious ways of 'spot-cooling' have been invented. One method provides a small 'pip' of glass drawn out from the surface of the tube, which, being away from the arc becomes cooler. Another system uses an electric junction (reverse Peltier Effect) to get the necessary cool spot. A widely used method which does not require any modification to the luminaire is to use a fluorescent tube fitted with a ring of indium (a rare metal) which, with certain other adjustments to the pressure and gas composition within the tube, enables it to operate efficiently at equatorial temperatures. This latter type of tube can be used successfully to replace ordinary standard tubes when the luminaire is in a hot environment, provided that the ambient temperature does not exceed the tolerance of the lamp, luminaire, control gear and wiring.

**10.1.5** While MBF lamps are little affected by ambient temperature (10.2.1), SON lamps, especially when operated in enclosed luminaires in high ambient temperatures can be unsatisfactory because of variation in arc voltage. Davies proposed a method of determining the acceptable range of conditions of operating these lamps[6], but for the purposes of this volume it will suffice to advise that where SON lamps are proposed for use in ambient temperatures outside normal human tolerance it will be wise to consult the manufacturer as to suitability of the particular combination of lamp, control-gear and luminaire in the expected temperature range.

## 10.2 Lighting in low ambient temperatures

**10.2.1** Most HID lamps will start satisfactorily and run up to full output in temperatures down to −5 °C (23 °F), and are more likely to be affected by damp and condensation than temperature. Tungsten-halogen lamps may give short life at temperatures below −5 °C if the lamps cannot get warm enough to operate their 'halogen cycle'. Ice forming on these lamps when they are not switched on can cause this problem, and they are also prone to burning of the lamp/lampholder contacts under these conditions. Fluorescent tubular lamps may give difficulty in starting at ambient temperatures below 5 °C, particularly under damp conditions, and will not give their full light output if the tube wall temperature is substantially below 38 °C. In unheated industrial buildings in winter, fluorescent tubes may give a little trouble in starting, and it will be found that 'Switch Start' luminaires usually give the best service in these conditions.

**10.2.2** In refrigerated cold-stores, fluorescent tubes are successfully operated at temperatures down to about −20 °C, by using fully-enclosed luminaires. To start the tubes initially they are warmed-up before being inserted in the luminaires, and they are then operated continuously to burn-out without switching. Under these conditions the tubes may give about two years continuous burning, perhaps more if there are no voltage fluctuations to cause re-striking. A useful tip is never to locate fluorescent luminaires close to the cold-air recirculators, for the cold blast can literally 'blow the lamps out' due to chilling. Lighting manufacturers can supply a special type of fluorescent tube in which the gas pressure is adjusted to aid low-temperature starting. Fully enclosed luminaires are preferred, and under extreme conditions the luminaire may contain a tungsten-filament lamp to act as a heater, or a small heater element may be incorporated into the luminaire. Because of possible problems, it is always wise to advise the lighting supplier of the intention to use lighting equipment at low ambient temperatures—they may offer variants of their standard products to suit the duty.

**10.2.3** It sometimes happens that, for convenience, the user employs the same type of luminaire as used in his interior lighting for such locations as under loading bay canopies, or in adjacent unheated garages and stores. In some cases this leads to problems, if not from low temperature starting then from corrosion or rust due to damp. Luminaires designed for interior use will not give long and safe service under these conditions (7.4.3).

## 10.3 Lighting for clean rooms and sterile rooms

**10.3.1** *Clean rooms* are used for industrial processes under conditions of stringent cleanliness. To produce a dust-free atmosphere, they are sealed rooms with 100 per cent controlled mechanical ventilation. In order to get very low air-speeds (and thus avoid entraining dust particles in the air-flow) input and output grilles of large area are used—in some cases consisting of virtually the whole floor and the whole ceiling. Such rooms produce 'laminar air-flow', and represent considerable capital investment. The work performed in such rooms is usually of meticulous accuracy, requiring illuminances of at least 1000 lux. The heat from the lamps is ducted away with the air-flow, but it is not good practice to use ordinary 'air-handling luminaires' (e.g. those in which the effluent air is drawn through over the tubes) as dust can be trapped in the luminaires or on the tubes. It is preferable to place the tubes behind glass or clear plastic panels, and to gain access to them from outside the controlled atmosphere enclosure (Figure 10.1). Clean rooms are used in the pharmaceutical industry (19.1.11), in fine engineering work and microelectronics (19.3) and the preparation of dust-sensitive items such as the drums used in xerography, componentry in nuclear research, etc.

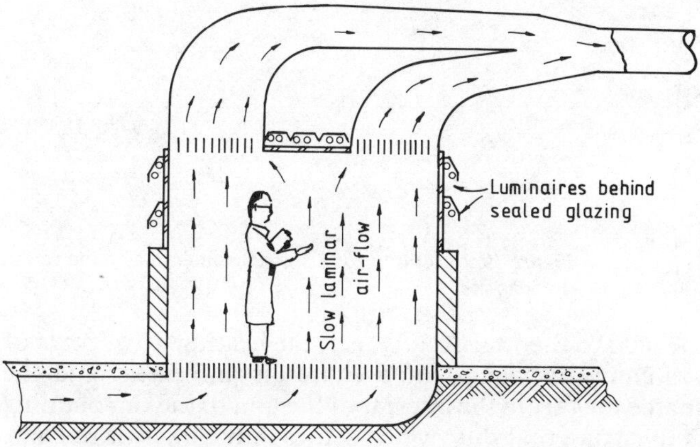


*Figure 10.1* Lighting for a laminar-flow clean room. Luminaires are set behind sealed glazed panels in the ceiling and walls as needed for the task. Air enters through the large grille which forms most of the floor, and leaves via similar large grilles forming most of the ceiling. The slow air movement ensures that there is minimum entrainment of dust.

**10.3.2** *Sterile rooms* present slightly different conditions to those met in *clean rooms* (10.3.1) in that in the latter the whole room has to be kept dust-free; while in the former, not only must the whole room be dust-free but all or parts must be of surgically sterile cleanliness. Sterile rooms are used for preparation and packing of pharmaceutical products and dressings etc., and usually must accommodate many operators. The general lighting and ventilation of the room may be arranged in similar manner to clean rooms practice, but additional enclosed or hooded benches are located in the area. Each bench or work area is provided with a positive pressure of sterilized air, so that contamination cannot be drawn into these critical spaces by air-flow from the general room volume. Because of the hooding or enclosing of work-benches, local lighting will be incorporated at each work-station, with a glass or clear plastic window separating the luminaires from the sterile zone (Figure 10.2).

**10.3.3** For all applications of lighting where very high standards of cleanliness are required, the following special guide-lines for the construction of fluorescent luminaires will be found helpful:

(a) The whole lighting unit shall be capable of withstanding repeated thorough cleansing with water and detergent and an agreed type of disinfecting chemical,

(b) Apertures giving access to lamp compartments, gear compartments or wiring chambers etc must be as small as possible to limit the movement of air between those inner chambers and the surrounding air. Spaces within louvres and light controllers shall be wide enough to permit easy cleaning.

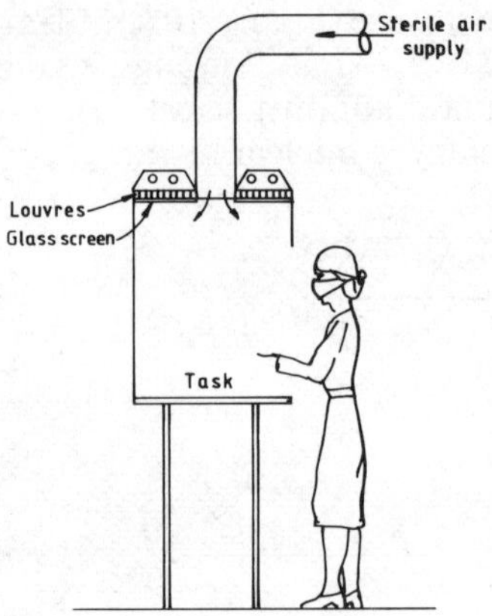


*Figure 10.2* Lighting for a sterile cabinet in a sterile room or clean room.

(c) Attention must be paid to thermal cycling, e.g. the heating up of parts of the luminaire on switching on which could cause air movement and the transfer of contaminated dust from the interior of the unit to the surrounding air. The objective is to ensure that dust which settles in the interior remains there, and does not get entrained in air-flow and transported out again.
(d) If a sealed fully enclosed luminaire is to be used in a sterile area, it must be fitted with a breather filled with a filter material fine enough to trap particles of a size to be agreed with the medical or technical experts. For a few very special applications, the sealed interior may be vented through water-bottle filters which will prevent both the entry and the egress of micro-organisms.
(f) Packing materials and gaskets must be selected from substances which will not support mould growth, and which are not hygroscopic.
(g) The preferred designs for luminaires for these special duties will be those having the minimum horizontal area upon which dust can settle.
(h) For very critical situations, consideration should be given to the use of luminaires that can be removed from the clean area periodically and taken to a cleansing area for sterilizing. In conditions of extreme difficulty where contaminated luminaires cannot be satisfactorily sterilized, they must be replaced. Those so contaminated in nuclear establishments must be disposed of in 'graves' (deep concrete-lined boreholes). Luminaires may be specially constructed so they are easy to dismantle for cleaning. Dust-tight luminaires to BS 4533[27] and provided with exterior surfaces and features that make them easier to clean may be used.

## 10.4 Lighting with reduced r.f. interference

**10.4.1** All gaseous discharge lamps (HID lamps and fluorescent tubes) produce emanations in the range of radio frequencies which can cause interference with communications equipment. The interference in the radiofrequency (r.f.) bands may be of various types:
(a) Mains-borne interference; if not adequately isolated by a r.f. filter, frequencies can leak into the mains and affect nearby sensitive equipment.
(b) Mains-borne and radiated interference; equipment which is not isolated by an r.f. filter may inject interference into the connecting cables, and, if

these cables are not adequately screened with earthed metal, the cables can radiate interference which can affect nearby sensitive equipment.
(c) Radiated interference; emanations may be radiated from lamps, control gear circuitry, and from ignitors for HID lamps. (Ignitors comprise circuits to produce high-voltage high-frequency pulses for starting lamps. Once the lamp starts, the ignitor is out of circuit; so interference due to this cause is usually of brief duration and confined to a few seconds when the lamp is attempting to 'strike'.)

**10.4.2** Interference can vary from 'clicks' (due to peaks and inverse-peaks generated during switching and starting of lamps) to continuous interference in the r.f. and audio frequencies of a regular or varying nature. It can affect sensitive equipment such as electrical test instruments, televisions and radios, security monitoring devices, computers, and recording and broadcasting apparatus. Interference from commonly-used types of lighting apparatus are most frequent on Long Wave and Medium Wave radio bands, and on Bands I and III, affecting domestic radio and television reception. The effects in industry can be serious, particularly now that there are so many applications of microprocessors and computer-controlled processes. The *'Wireless Telegraphy (Control of Interference from Fluorescent Lighting Apparatus) Regulations 1978'* came into force at the beginning of 1979, prompting the Lighting Industry Federation to issue sets of *Guidance Notes to Manufacturers* on how to comply with the *Regulations.* Radio interference is covered in BS 4533[(27)], and also by BS 5394: Part 1: 1976[(34)].

**10.4.3** Fluorescent tubular lamps operating on dimmer circuits (both variable dimmers and those giving pre-set outputs), and all such lamps operating on solid-state high-frequency control-gears (as do most emergency lighting luminaires (6.4)) are prone to producing r.f. interference. It is quite practical to suppress the mains-borne emanations[(35)]. In cases of complaint against interference by others, enquiries may be made to The Home Office, Directorate of Radio Technology, Radio Interference Branch, Waterloo Bridge House, Waterloo Road, London, SE1 8UA. The use of dimmers which use 'wave-chopping' (e.g. thyrister dimmers) gives rise to waveform distortion, and this subject of harmonics in mains supplies is covered by an available publication[(36)].

**10.4.4** In broadcasting and recording studios, and laboratories where delicate electrostatic equipment is used, the problem may be radiated emanations from fluorescent tubes. Care must be exercised to ensure that all live metal is enclosed within earthed metallic screens. Radiation from the tubes themselves can be very substantially attenuated by sleeving the tubes in a fine metallic mesh which is earthed, or fitting such an earthed mesh over the non-metallic parts of the luminaire enclosure. It will be found that lighting manufacturers are sympathetic to the difficulties that users may have in respect of interference, but, to gain their collaboration in solving the problems it is vital to consult them at the outset, and not wait until a new installation is giving trouble due to radiointerference, when it may be costly, if not impossible to correct the problem without re-equipping and re-wiring.

Chapter 11

# Lighting in hostile environments

This chapter continues the theme of Chapter 10, on matching the lighting to the environmental conditions at the place of use. It also warns the user about the types of abuse of the lighting installation which may occur under conditions of poor supervision and management (11.3). On some large factory sites, conditions such as are discussed in this chapter and in Chapters 10 and 12 may all exist in different buildings.

## 11.1 Lighting in dusty or soiled atmospheres

**11.1.1** Light which is obstructed by dirt on the lamp and on the luminaire surfaces still has to be paid for, though it cannot benefit the user. An allowance is made in calculations for there to be a certain degree of acceptable light loss (16.2), and it is the recovery of light which would otherwise be lost that is the justification for the cost of maintenance procedures (13.4). In addition to diminishing the light output of the luminaire, a layer of dust or other pollution will tend to act as diffusing layer, with the result that the luminaire will not achieve its intended light distribution. In general, the beam angle of luminaires is widened, and the peak diminished by a layer of dirt; and the luminaire may as a result project more light at angles which were previously optically shielded, so that the direct glare from the luminaire is increased. An increase in glare diminishes the 'seeing value' of the remaining illuminance, so a satisfactory lighting installation can become thoroughly unsatisfactory in a short time through simple neglect.

**11.1.2** It is apparent that considerable benefit can be gained by carefully specifying the type of luminaire (open, through-vented, enclosed, etc) according the nature and degree of atmospheric pollution which is expected at the place of use. Constructions and performance of various types of luminaires for industrial use are covered by BS 4533[27], and the user's preferences in terms of that Standard are fundamental to the success (in function and cost) of the installation, and should form part of the Outline Lighting Specification (7.1). Considerations of the environment in which the equipment will operate are important to choice of lighting equipment (8.1).

**11.1.3** Even luminaires designed as 'dust-tight'—e.g. totally enclosed to BS 4533[27], will tend to 'breathe', expelling heated air as they warm up, and drawing in air (and contaminants) as they cool down—unless they are fitted with a breather-filter of suitable design (10.3.3). Over a period of time, through a succession of such thermal cycles, quite large quantities of contaminants may be drawn into the interior of an enclosed luminaire. The results will depend on the nature of the foreign substance; e.g. conductive dusts (such as carbon dust, finely divided metallic dusts etc) can cause tracking and electrical breakdown of the luminaire; sooty, oily particles can rapidly occlude the light from luminaire; organic dusts (such as flour, spices etc) are usually hygroscopic, and will support mould growth. It is not unknown for a luminaire to start filling up with water after being contaminated in the latter fashion.

**11.1.4** Luminaires operating in atmospheres contaminated with fibrous dusts (e.g. wood-dust, cotton, wool, asbestos etc) can become so coated with air-borne particles that the settled material forms a thermally insulating layer on the luminaire and can cause it to overheat. Fats and oils, passing into the atmosphere in fine droplet form can condense out on a cold luminaire, and later drip off when heated as the luminaire is switched—and this can cause contamination of foodstuffs (19.1). Many other examples could be quoted of contamination of luminaires by dusts and air-borne soiling, giving rise to worsened lighting conditions or failure of the lighting equipment.

**11.1.5** The problems due to dusty and soiled atmospheres can be combatted by specifying an appropriate type of luminaire, for which BS 4533[27] is the most useful reference. Placing a translucent covering over the lamp is not necessarily the best way of reducing light-loss, for, unless the luminaire is properly enclosed and sealed, one will have interposed more surfaces between the lamp and the working plane, and the accretion of dirt on the reflecting and transmitting surfaces of the luminaire will therefore have increased effect in diminishing nett light output (Figure 11.1).

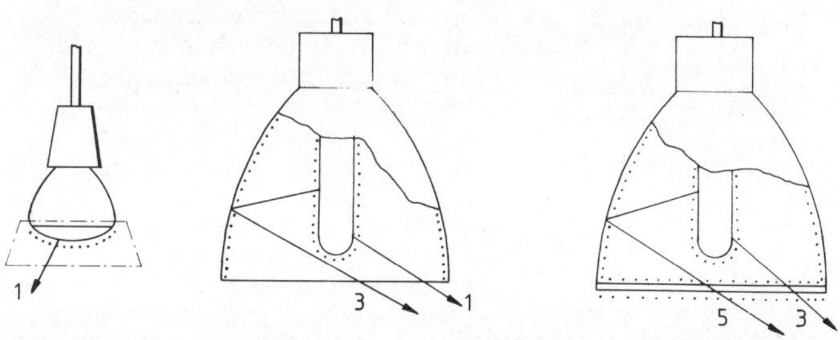


*Figure 11.1* Effect of soiling of luminaries. The numbers on the arrows indicate the number of layers of dirt through which an incident ray has to pass. (a) Reflector-lamp; (b) Tubular or isothermal lamp in an open reflector; (c) Tubular or isothermal lamp in an enclosed luminaire which is not properly sealed. Of these (a) gives the best performance in relation to accreted dirt, and (c) gives the worst. But, if (c) can be properly sealed, it may give performance equal to or better than (a). The dotted surfaces are those which may acquire a layer of dirt.

**11.1.6** Where very frequent cleansing of luminaires is necessitated by heavy pollution, it is sometimes possible to keep the luminaires reasonably clean by jetting them with water, but certain precautions are necessary (11.2.1). Most kinds of dirt particles can only settle on surfaces if the speed of the air in which they are entrained is below a critical speed. Thus, if it can be arranged that the surface which it is desired to keep clean is subjected to an air-current moving at faster than the maximum speed which settling can take place, then the surface will remain clean. This is the theory behind the construction of *through-vented luminaires.* There are two main types: the more common one in which air passes through the luminaire (11.1.7), and another kind in which air is conducted through a path exterior to the luminaire enclosure (11.1.8).

**11.1.7** The through-venting principle is applied to luminaires for both fluorescent tubular lamps and HID lamps. Slotted-top trough reflectors for fluorescent tubular lamps have been around since the very early days of this light-source, when it was found that permitting a small amount of upward light removed the *tunnel effect* (a gloomy and rather depressing appearance to an interior due to the ceiling receiving little light, and creating an illusion that there is a dark ceiling immediately above the level of the luminaires). It was noticed that slotted-top trough luminaires tended to keep cleaner than those with solid tops, and eventually slotted-top luminaires were devised in which the construction facilitated the air-flow and improved the self-cleaning effect (Figure 11.2 (a)). The principle of through-venting to get the self-cleaning effect is also applied to some luminaires for HID lamps, where another benefit occurs, namely that the convection of air through the luminaire can be utilized to cool the gear-compartment. This may not only extend the life or increase the reliability of the control-gear, but it can permit the control-gear compartment to be smaller and closer to the reflector—resulting in a less costly design of luminaire (Figure 11.2 (b)).

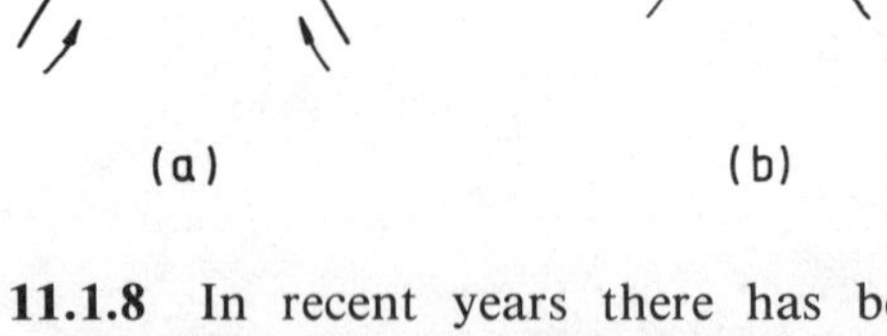


*Figure 11.2* Through-vented luminaires. (a) Typical cross-section of slotted-top fluorescent-tube trough luminaire. The slots permit passage of air to keep the luminaire reflecting surfaces and lamps clean, and also allow some 5 to 10 per cent of the lamp lumens to escape upward. (b) Typical cross-section of a HID-lamp luminaire having the through-venting principle to get the self-cleaning effect and to cool the control-gear compartment.

**11.1.8** In recent years there has been a further development in the application of through-venting, with the development of a type of high-bay luminaire in which the path for the convected current of air passes outside the luminaire enclosure. This can be arranged so that the air drawn into the convection path passes across the glass panel enclosing the base of the luminaire, the air current exerting a scouring action on the glass surface and thus keeping it cleaner than most other kinds of high-bay luminaires, even under conditions of very heavy pollution (Figure 11.3).

**11.1.9** When discussing the suitability of luminaires for use in dusty atmospheres, the fire risks associated with suspensions of dusts should not be overlooked. Finely divided materials such as flour, custard powder, cornflour, and even icing sugar, can burn violently if suspended in the atmosphere. If the density of particles is such that each particle is within the

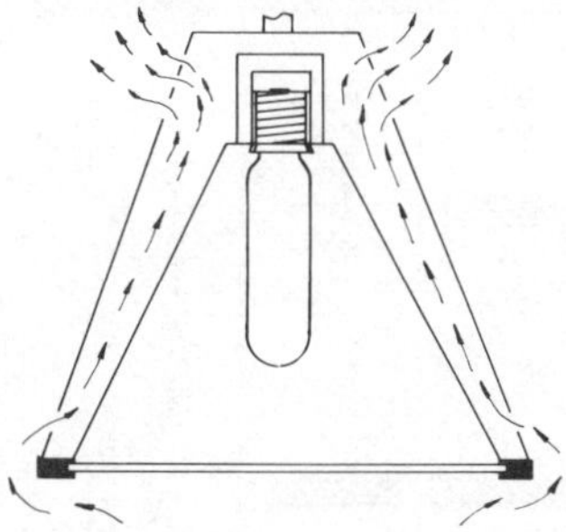

*Figure 11.3* Through-vented luminaire with an external air-path. The air current cools the lampholder and deflects the flow of hot air away from the gear compartment above the luminaire (not shown). The air flow also tends to scour the face of the enclosing glass plate and helps keep it cleaner. (Simplified sketch of the 'Nimbus Industrial Downliter'. Courtney, Pope Lighting Ltd.)

flame zone of an adjacent particle, once conflagration commences, the flame may be propagated with great speed and violence to cause destructive explosion—of sufficient power to demolish buildings. This possibility should be considered in discussions with the fire prevention officer and others concerned, and a suitable specification for the lighting equipment decided (19.1). The important factors will be the expected degree of penetration into the luminaire enclosures (determined by the size of particles and the luminaire construction) and the temperature at which the material under consideration will combust. Luminaires which are designed to give a controlled external temperature, e.g. as those designed for Zone 2 (12.2), may be suitable in some cases; indeed, such luminaires when provided with a corrosion-resistant finish may be almost universally applicable in dusty, soiled, wet and corrosive atmospheres where there is no Zone 1 or Zone 2 requirement (12.1).

**11.1.10** A type of luminaire which is discussed again in Chapter 12 is the *pressurized* luminaire, which is an enclosed type of luminaire in which the internal pressure is kept above atmospheric pressure by connection to an air-line. There is usually a purging valve so that there is a small continuous flow of clean air through the luminaire, which may have a cooling effect on the luminaire as well as keeping it clean. These luminaires have applications in the food and pharmaceutical industries (19.1). With adaptations, they may be made suitable to use in hazardous Zones (12.2).

**11.1.11** Reference should be made to BS 4533[27] regarding dust-tight and dust-proof constructions for luminaires. Degrees of resistance to penetration of foreign bodies and dusts are listed in Specification UTE C 20 010 (and in BS 4533) as shown in Table 8. Thus, complete protection against dust would be designated 'IP 6 - -'.

## 11.2 Lighting in wet and corrosive atmospheres

**11.2.1** The constructional differences between *dust-tight* luminaires (11.1) and *water-jet-proof* luminaires to BS 4533[27] are not great, the most important difference being that the casing of the luminaire must be corrosion-resistant. Non-metallic luminaires, constructed entirely of plastics or glass-fibre reinforced plastics can be highly effective for this duty. Alternatively, luminaires constructed from metal castings, pressings or spinnings, which have been coated with a suitable plastic covering can be equally suitable. Water-jet-proof luminaires, in addition to service in wet

1st figure: protection against solid bodies

| IP | P | tests | |
|---|---|---|---|
| 0 | 0 | | No protection |
| 1 | 1 | 52·5mm | Protected against solid bodies larger than 50 mm (eg: accidental contact with the hand) |
| 2 | 2 | 12·5mm | Protected against solid bodies larger than 12mm (eg: finger of the hand) |
| 3 | - | ϕ2·5mm | Protected against solid bodies larger than 2·5mm (eg: tools, wires) |
| 4 | 3 | ϕ1mm | Protected against solid bodies larger than 1mm (fine tools and small wires) |
| 5 | 4 | | Protected against dust (no harmful deposit) |
| 6 | 5 | | Completely protected against dust |

The first two figures are defined in identical fashion by the standards U.T.E. C 20 010, C.E.I. 144 and DIN 40 050 P DIN 40 050 (1963)

2nd figure: protection against liquids

| IP | P | tests | |
|---|---|---|---|
| 0 | 0 | | No protection |
| 1 | 1 | | Protected against vertically-falling drops of water (condensation) |
| 2 | 1S | 15° | Protected against drops of water falling at up to 15° from the vertical |
| 3 | 2 | 60° | Protected against drops of rain water at up to 60° from the vertical |
| 4 | 3 | | Protected against projections of water from all directions |
| 5 | 4 | | Protected against jets of water from hosenozzle from all directions |
| 6 | 4S | | Protected against jets of water of similar force to heavy seas |
| 7 | 5HI | 1m; 15cm clear | Protected against the effects of immersion |

3rd figure: mechanical protection

| IP | tests | |
|---|---|---|
| 0 | | No protection |
| 1 | 150g; 15cm | Impact energy 0.225 joules |
| 2 | 250g; 15cm | Impact energy 0.375 joules |
| 3 | 250g; 20cm | Impact energy 0.500 joules |
| 5 | 500g; 40 cm | Impact energy 2.00 joules |
| 7 | 15 kg; 40 cm | Impact energy 6.00 joules |
| 9 | 5 kg; 40 cm | Impact energy 20.00 joules |

The third figure is defined by the French standard U.T.E. C 20 010 it is under international study at the C.E.E. - C.E.I.

**Table 8** Degree of protection of casings of electrical material up to 1000 V a.c. and 1500 V d.c. (the IP System)

places and places where they may need to be hosed for hygienic reasons (e.g. in abbatoirs), are also suitable for use in dusty atmospheres, provided there is no objection to the use of water in that environment (11.1.6); however, it is vital to note that the practice of cleaning luminaires with a water jet is potentially highly dangerous if the installation contains any luminaires which are not water-jet-proof. If an unsuitable luminaire was jetted, the operator could be in danger of electrocution; the equipment would also probably be damaged.

**11.2.2** The user should be careful to distinguish between classifications of BS 4533[27]; a *drip-proof* enclosure will not necessarily be rain-proof; a *rain-proof* enclosure will not necessarily be water-jet-proof; a water-*jet-proof* enclosure will almost certainly not be watertight; a *watertight* enclosure may not be suitable for submersion; finally, *submersible* enclosures are not necessarily water-tight, but may be constructed on the diving-bell principle.
**11.2.3** There are a few industrial environments in which the atmosphere contains entrained droplets of corrosive substances, such as in wool-stripping, hide pickling and leathermaking; other atmospheres may contain steam or droplets of substances such as acetic acid or vinegar, a common situation in the food industry. In all these places where there is wetness or a corrosive atmosphere, it is not only the luminaire which must withstand the hostile environment; the entire electrical installation is required to present resistance to the conditions to ensure electrical safety.
**11.2.4** There is a potential danger to the operator when it is necessary to open up enclosed luminaires for the purpose of cleaning or re-lamping. If the atmospheric conditions are adverse at the time, it is essential to isolate the luminaires from the mains before opening them. Indeed, it would be better practice to time such maintenance operations so they may be carried out when the plant is shut down.
**11.2.5** Reference should be made to BS 4533[27] regarding protection of constructions against penetration of water and other liquids. Degrees of resistance to penetration are listed in Specification UTE C 20 010 (and in BS 4533) as shown in Figure 11.4. Thus, 'protection against jets of water from hosenozzle from all directions' would be designated 'IP - 5 -'.

## 11.3 Lighting for rugged environments

**11.3.1** In general, Chapters 10, 11, and 12 deal with the methods of matching the lighting equipment to the special environments met in various kinds of industrial buildings. Were human nature perfect, and if all who are employed in industry always performed their work with care and without moments of thoughtlessness or irresponsibility, it would not be necessary to include this Section 11.3 which deals with the abuse of lighting and electrical equipment in industrial installations. The reader is assured that all the types of abuse mentioned in this Section actually do occur; the author witnessed the acts or their results in the incidents described (11.3.2).
**11.3.2** The following are just some of examples of the kinds of abuse and mis-use of equipment that are known to occur, many of which can result in costly damage, serious risk to personnel, or may have other unfortunate results. The brief description of each will prompt the reader to his own solution of what might be done to prevent the abuse or to minimise its potentially adverse results:
(a) Suspending a chain-hoist from overhead lighting trunking to lift a heavy load.
(b) Using a bayonet-cap adaptor to take an improvised electric supply to feed an un-earthed handlamp.
(c) Using a wall-mounted switch as a foot-hold while climbing to reach a high object.

(d) Placing a number of heavy carpet strips over a low-mounted suspended fluorescent trough luminaire to keep them off the floor while carrying out repairs.
(e) Standing on the tines of a fork-lift truck and getting the operator to elevate the lift in order to get access to replace an MBF lamp mounted 6 m above the floor.
(f) Attempting to remove the remains of a smashed MBFR lamp from its socket by wrapping a piece of cloth round it, while the circuit was still live.
(g) Inspecting the inside of metal vat (e.g. an 'earthy location') by a fluorescent batten luminaire connected to a 13-ampere socket by means of bare cable-ends pushed into the socket, the shutter being held aside with a screwdriver (earth wire not used).
(h) Unofficially experimenting to find out if a 400 W MBF lamp would work in a luminaire intended for a 400 W SON lamp.
(i) Moving a fluorescent luminaire to a new position, suspending it on two pieces of string, and extending the connecting cable with a piece of lighting flex—the connections made by twisting the bare ends which were left exposed.
(j) Unhooking a high-bay MBF luminaire from its suspension point, and allowing it to hang down on its pvc/pvc connecting cable.
(k) Placing a pole across two suspended luminaires and using this to support some wet canvas sheets to dry overnight.
(l) Connecting a festoon of Christmas-tree lights into a wall-switch on a staircase, and leaving the cover of the switch off.
(m) Attempting to liquify some petroleum jelly by immersing a 100 W tungsten-filament lamp in it.

All the 'unofficial modifications' to lighting equipment mentioned here contravene the *IEE Regulations*(15), and all these occurrences contravene the *Health & Safety at Work Etc Act*(26) and could be the subject of prosecutions against the persons concerned and the companies on whose premises the offences occurred. It is thus apparent that for environments where there may be abuses, lighting equipment must be selected with care for its duties, robustly installed and frequently inspected. If the installation is suitable, and is properly modified to suit locally changed needs when they occur, there should be little motivation for unofficial interference with the lighting installation.

**11.3.3** Reference should be made to BS 4533(27) regarding enclosures having suitable resistance to mechanical force. Degrees of protection are listed in the French Standard UTE C 20 010, and this provides the third figure in the IP description; thus, mechanical protection to withstand an impact of 20 joules would be designated 'IP - -9' (Figure 11.4).

## 11.4 Lighting in windy or vibrating environments

**11.4.1** Suspended luminaires in industrial high-roof buildings may be subjected to considerable wind-pressures when large doors are opened, sufficient to fracture rigid suspension drops (8.2.5). If swinging rigid suspensions or chain suspensions are employed in situations where luminaires may be subjected to strong internal air movements, then lateral staying with rods or cables may be necessary. A better method is to mount the luminaires on rigid overhead trunking (8.2.2).

**11.4.2** Steel-framed skin-cladded industrial buildings may be subject to considerable and sustained vibration due to wind pressures or from the operation of a gantry-crane. Mechanical shocks can be sufficient to seriously shorten lamp life or to cause lamps to fracture in their sockets. Under conditions of extreme vibration, fluorescent tubular lamps have been known to shear their connecting pins in the lampholders and to fall out of open luminaires. Screw-cap lamps, if not firmly screwed home, can fall from their sockets under severe vibration. The first approach to dealing with such problems should be to attack the source of the vibration, e.g. any slackness in the bolting up of the structure or its gantry etc, and putting soft bumper stops on large sliding doors. Next, the method of attachment of the lighting equipment to the structure should be examined to ensure that all weight is taken by solid members of the roof and not by flimsy subsidiary members which were not designed for the purpose and which therefore flex excessively under stress. Consideration should be given to interposing some heavier support steelwork between the structure and the luminaires, and in bad cases advice should be sought on the use of anti-vibration mounts. Note that the wrong design of anti-vibration mount can actually amplify the vibration. In conditions of unavoidable vibration or mechanical shock, select luminaires which incorporate robust lamp-steadies, and consider using enclosed luminaires fitted with wire guards, or using enclosed luminaires to prevent a detached lamp or parts of a fractured lamp from falling. (This is a requirement of law in the case of food factories on grounds of hygiene (19.1), and is a general requirement under Health and Safety Regulations (Appendix I) for all industrial buildings where there is hazard from vibration or mechanical shock.)

Chapter 12

# Lighting in flame-hazard environments

This chapter reviews the salient points about provision of lighting in areas where there is a fire or explosion due to the presence of flammable materials.† This is a topic under which the user should not hesitate to take professional advice in view of the complexity of the technical and legal requirements. Lighting equipment for these areas is costly, for it is constructed to stringent standards; but, with imaginative design it is sometimes possible to reduce costs substantially (12.3). The problems which can arise during installation (12.4) should not be glossed over; the guide-lines given arise from practical experience rather than from theoretical considerations.

## 12.1 Zone classifications; occupier's responsibilities

**12.1.1** Areas where there may be a hazard due to the presence of flammable liquids, vapours, gases, fumes etc, are classified according to the degree of hazard into three types which are designated as 'Zones'* in IEC Standard 79-10, thus:

*Zone 0:* An area where an explosive gas-air mixture is continuously present or is present for long periods.

*Zone 1:* An area where an explosive gas-air mixture is likely to occur during normal operation of the premises.

*Zone 2:* An area in which flammable and explosive substances are so well under conditions of control that a hazard is likely only under abnormal conditions.

It is sometimes difficult to determine where the border of one Zone ends and that of another begins, and as it is the occupier's responsibility to ensure that only appropriate *'proof'* luminaires and electrical equipment are used in each Zone, this is a matter upon which the occupier will probably need to take advice from the relevant authorities, e.g. in the UK from the Health & Safety Executive Inspector. Note that the Zones are conceived in the vertical as well as in the horizontal plane; thus, in an area where there might be heavy petroleum vapour, while only proof fittings could be used within a certain

†Note the correct use of the term *'flammable'*, meaning 'easy to set on fire'; the traditional word 'inflammable' means exactly the same thing, but its prefix 'in-' suggests the opposite, particularly to non-English speakers, and is no longer used in British Standards.

*Previously, these were termed 'Divisions'.

distance above the ground, above that level equipment of lesser protection might be used. In such a case, the Zone 0 might apply to any pits, depressions in the ground, and spaces under vessels, from the lowest level up to the height above which it is judged that the gas-air mixture would be unlikely to remain for long periods. Above this might be a further strata designated as Zone 1, and yet another above that designated Zone 2 (Figure 12.1). Areas (horizontally and vertically defined) which are not allocated as either Zones 0, 1, or 2, are simply termed 'normal atmospheres' (i.e. the use of the unofficial term 'Zone 3' is deprecated).

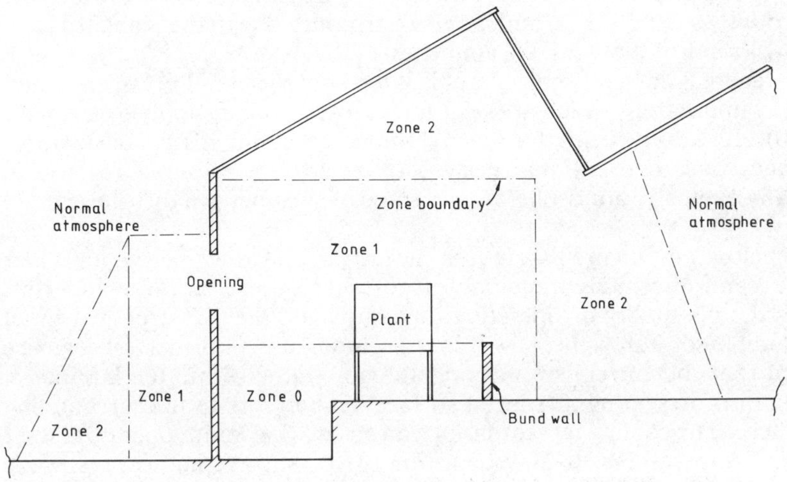


*Figure 12.1* Hazardous zones. The sketch shows a building in cross-section (not to scale), indicating how Zones might be allotted in respect of risks associated with plant emitting a flammable vapour that is heavier than air.

## 12.2 Classification of protected equipment

**12.2.1** Luminaires and electrical equipment for use in Hazardous Zones (12.1) must be protected to prevent fire or explosion being caused by their temperature or by the formation of sparks. Various kinds of protection is possible:

(a) *intrinsically-safe* equipment may be used, viz equipment so constructed that it cannot generate a spark or temperature high enough to cause ignition of the flammable gas-air mixture;

(b) electrical apparatus that is not intrinsically-safe may be enclosed in a *flameproof enclosure*. Flameproof construction is a means of housing electrical apparatus which, in normal operation, does not produce sparks or temperatures above the ignition temperature of the surrounding explosive gas–air mixture, but the construction of the enclosure is such as to withstand the force of an explosion within itself from this cause, without propagating flame or heated gases outside itself. The method of construction is on the 'gap and flange' principle, in which wide flanges, accurately machined flat, are juxtaposed at the joints between the parts of the enclosure, the gap between them being small enough and long enough to cool the gases to a safe temperature before they escape to the surrounding atmosphere.

(c) other methods of protection may be employed, e.g. segregation, ventilation and *pressurization* (12.2.3).

(d) for Zone 2 areas, safety may be achieved by the selection, use and installation of electrical apparatus, viz luminaires may be used in which generation of sparks is prevented, and in which the temperature rise in use is restricted.

**12.2.2** The Code of Practice for the selection, installation and maintenance of electrical apparatus for use in potentially explosive atmospheres for industry is BS 5345[38] which is being published in parts. At the time of writing, only Parts 1, 3, 4 and 6 are available. This Standard covers much the same ground as CP 1003,[39] and is being harmonized with the Standards and Codes of Practice of other EEC countries.

**12.2.3** There is currently considerable interest in the use of pressurized and ventilated luminaires, which are continuously fed with a supply of clean air (11.1.10). It would seem that such luminaires could offer considerable advantages over conventional designs of enclosures required for use in potentially explosive atmospheres, and some of the points in their favour are as follows:

(a) An enclosed luminaire that is continuously supplied with clean air under pressure cannot aspirate flammable substances nor moisture, and thus presents the possibility of operating the lamps and their control gear in an uncontaminated atmosphere within the luminaire. Further, it can be arranged that the current of air cools the apparatus within the luminaire, resulting in it presenting a reduced surface temperature to the surrounding atmosphere. The cooling by ventilation can ensure that lamps operate at their optimum temperatures. It may be arranged so that as long as pressure is applied from a local air receiver, compressor or air ring-main, there would be a flow of effluent air out of the luminaire enclosure. While the air pressure is so applied, an air-pressure-actuated contactor in a safe zone will maintain the electrical supply to the luminaire, but any diminution of the air supply pressure would cause the luminaire to become electrically isolated. This would protect against any possibly dangerous consequences of opening the luminaire in the presence of flame hazardous atmospheric conditions.

(b) Because such a pressurized luminaire does not have to be constructed to withstand the force of an explosion within itself, it may be constructed lightly and at much less cost than conventional Flameproof luminaires, thus relieving the economic constraint of the high cost of conventional Flameproof installations which often results in the provision of poor lighting standards in places where better lighting could possibly make a contribution to greater safety or productivity.

(c) Pressurized luminaires as described may be provided with a non-sparking means of connection and disconnection at a local plug and socket, and the air pressure supply line fitted with a coupler which will permit connection and disconnection without loss of air pressure. Thus it would be possible to enter a contaminated hazard zone and in complete safety remove any luminaire so that it may be replaced with another. Removed luminaires would be serviced in a normal atmosphere and returned as required to the installation. This provides a means of performing routine servicing of lighting installations in continuously hazardous locations. Because of the lightweight construction (see (b)), the physical task of removing and replacing the luminaires would be made easier.

It must be commented that pressurized fluorescent luminaires have been successfully used in many installations in the UK, usually in locations having a hazard from petroleum vapours, e.g. garage and petroleum premises, and in storage areas for printing inks, and such installations have been duly approved by Petroleum Officers, HSE inspectors and insurers. What is advocated here is the extension of the use of this system of providing safety in lighting installations for hazardous Zones. Further, if a range of such luminaires, together with their compressors etc could be made commercially available, then the same kind of apparatus could be used in damp and dusty atmospheres where the air pressure system would keep the interiors of the luminaires dry and clean. It would seem that the pressurized system would be suitable for use in atmospheres containing difficult substances (such as hydrogen gas and corrosive gases), for all kinds of flammable and conductive dusts, indeed for all adverse and hostile atmospheres; and its general approval and adoption for all such circumstances would bring the boon of lowered costs of lighting to a wide range of situations where at present cumbersome and costly explosion-proof types of luminaires or other expensive equipment must be employed. In closing this section it should be stated that no electrical apparatus should be used in any potentially explosive or flammable atmosphere unless it has been approved for use in that atmosphere(38, 39).

## 12.3 Lighting design considerations

**12.3.1** Because of the relatively high cost of protected luminaires (12.2) for use in hazardous zones (12.1), it is well worth while giving some careful thought to the planning of departments which are to be designated as hazardous zones, to try and find ways of (a) making the designated hazard zone smaller, and (b) removing from the hazard zone all processes which are not in themselves hazardous. Further; if, for example, there is one hazardous process to be performed in a department, it may be less costly to remove that process to a separate smaller area, and thus avoid having to provide proof luminaires for the whole department.

**12.3.2** Nothing written here should be construed as advising users to 'bend the rules' relating to hazardous zones; there are, however, one or two procedures which, though unconventional, will in some cases enable the cost of the lighting installation to be substantially reduced. As an example of this approach, consider the well-known practice in exterior lighting of throwing light into a hazard zone from an adjacent area of normal atmosphere, a method that enables the lighting to be achieved using non-protected luminaires of good efficiency(1). A similar approach to the problem of lighting buildings having roof-lights could be to throw lighting into the interior from luminaires mounted in the open air above the roof glazing. The installation must be properly engineered to a standard approved by the Health & Safety Executive, the enforcing officer for the Fire Precautions Act (6.1) and the insurers, and it would be required that the roof glazing was sealed, and that the air extracted from the building did not pass near the luminaires. Access to the luminaires could be facilitated by the provision of walkways on the roof. Light losses through the glazing will be minimized if it can be arranged that the light flow is substantially normal to the plane of the

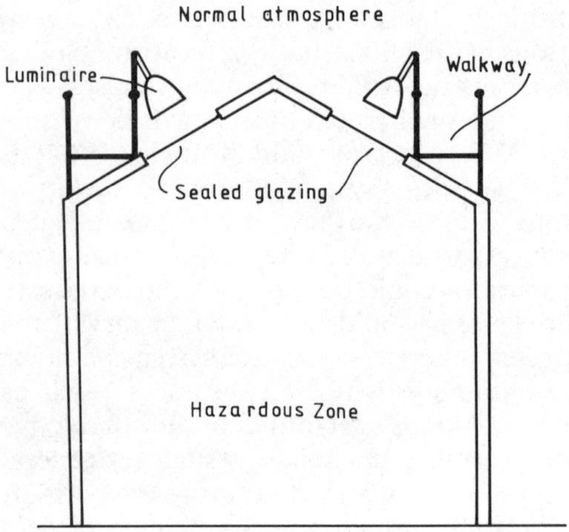


*Figure 12.2* Luminaires designed for exterior use in normal atmospheres mounted above sealed glazing to light an interior which is a hazardous zone. (Not to scale.)

glass, though there will inevitably be light loss if the glass is wire-reinforced and of substantial thickness (Figure 12.2).

## 12.4 Problems during installation

**12.4.1** There are few problems in regard to installing proof luminaires and electrical equipment in new premises, or in those where the plant is not yet in operation; but if the area of the installation is already a hazardous zone, the difficulties may be considerable. For example, it will be necessary to ensure that no sparks are created, and, of course, there must be no naked flames and no welding. This will require the use of nut-and-bolt fixings rather than welding supports to the structure, and usually the conditions will preclude the use of masonry drills and cartridge-operated bolt-fixers. Powered means of access and transport will have to be fitted with proof type electrical drives; and if engine-driven plant is needed, the engines will have to be of types adapted to and certified as safe to use in the presence of the particular atmospheric contaminant. No 'live' connections should be made in the hazardous zone unless the plant is shut down to safety, the area purged and ventilated, and a 'permit-to-work' issued by the responsible engineer. During the time the 'permit-to-work' is valid for electrical connections or hot processes etc, the permit for the normal operation of the plant is withdrawn, and only becomes effective again when the responsible engineer has carried out acceptance inspection of the new installation work and is satisfied that all has been left safe.

Chapter 13

# Maintenance of lighting installations

Difference in the cost of maintenance between one lighting proposal and another may be great enough to influence the choice of lighting system, and valuable savings may be made by correctly applying a Maintenance Factor in calculations, and determining the Economic Re-lamping Period. Specifying the preventative maintenance routines is part of the process of lighting design, and early choice of the mobile access equipment is advised as this may affect plant layouts (13.1). Often, it will prove more economic to use built-in access equipment, to enable all maintenance operations to be carried out at lowest cost and in greater safety (13.2). Money spent on regular preventative mainenance and re-lamping is not wasted; quite small expenditure will permit the performance of a maintenance programme that will result in a safe, clean and efficient lighting system and thus produce the lighting planned for at lowest cost-in-use and with extended technical life (14.4). As the care of the lighting installation is not usually the responsibility of the designer or specifier, too often the value of efficient preventative maintenance is underestimated (13.3). Great care must be exercised when considering relamping luminaires with lamps of different type or manufacturer to that specified in the original installation (13.4).

## 13.1 Designing for safe, low-cost maintenance

**13.1.1** Consideration of how the lighting installation will be maintained may be so important as to influence the choice of lighting equipment. For example, more costly enclosed luminaires which can be hosed clean from the ground might be justified compared with unenclosed luminaires which, though cheaper in capital cost, would be more expensive to own (11.2). Thought should be given at the outset as to how access to the roof or ceiling will be obtained during installation (7.4.1), and also for care during the life of the installation. All overhead lighting requires means of access for periodic re-lamping, cleaning and preventative or corrective maintenance. Solving the access problem is part of the lighting design, not a separate problem to be tackled later. It will involve a visit to site in the case of existing buildings, or, in the case of new buildings, careful study of the building plans and elevations and the plant layout. It will then be possible to decide whether to employ *mobile access equipment* or *built-in access equipment* (13.2).

**13.1.2** In planning access to high-mounted luminaires, it should be remembered that it may be difficult to bring mobile access equipment into position if there is fixed plant or other permanent obstruction at floor level; and that overhead pipes, trunking or cranes etc may impede access. Also, in foundries, papermills and other factories which use hot processes, it may be too uncomfortable or too dangerous for men to get up to the luminaires when the plant is working (10.1). Mobile access equipment (such as tower-ladders, hydraulic access platforms etc) will require space for movement, and this may influence the sizing of gangways between plant. A decision to build-in the means of access should preferably be made before the steelwork drawings for a new structure are completed; then, roof walkways and hatches, roof galleries and similar features can be incorporated at lowest possible cost.
**13.1.3** In estimating the total installed cost of a new lighting installation, and in calculating the total annual running cost (18.2), the cost of the access equipment should be included in the capital cost of the installation. It would be good accounting also to provide for a notional value for the floor area that becomes unusable for production purposes because of the need to provide extra-wide lanes between plant to permit the access equipment to pass through. Thus, some extra cost for access equipment which did not have this requirement could be justified, e.g. equipment which folds to narrow width for movement.
**13.1.4** All lighting installations require maintenance, periodic lamp replacement, cleaning, and some preventative maintenance (13.3) in order to maintain the illuminance at the required level with reliability, and to minimise the need for corrective maintenance. It can never be economic to 'over light' an area and allow the lighting level to fall by neglect; if a lesser illuminance will suffice, it must be cheaper to design for it and carry out a simple programme of maintenance (which will save energy as well as expense) (18.2). The provision of suitable means of access will simplify all relamping and maintenance procedures, resulting in a lower labour cost and reduction of danger to those carrying out the work. Improvisations, and hasty interventions by the maintenance department in response to complaints about lamp failures and inadequate lighting are evidence of a failure to manage the lighting well. Good management of lighting starts at the design brief (7.1) when the plan for maintenance will be considered as one of the factors of design and the choice of equipment. One of the elementary preparations for future good maintenance will be to compile a schedule of the lamps that will be required for replacement purposes later on, and to keep a copy where it is sure to be found, for example in a sealed plastic envelope attached to the switchgear. Modern HID lamps last a long time, and it can be five or six years between relampings, by which time memory cannot be relied upon, records may be lost, and the original personnel departed. There is considerable risk in inserting a lamp of a different type, or from a different maker than was specified originally for use with the luminaires (13.4).

## 13.2 Mobile and built-in access equipment

**13.2.1** This Section briefly reviews the main types of mobile access equipment which are used to gain access to luminaires, but is not intended to provide accurate descriptions of particular manufacturers' products. The maximum heights suggested as being the ranges of the equipments are indicative only, and particular manufacturers' products may have higher or lower reaches. The products reviewed are step-ladders (13.2.2); fixed and extending ladders (13.2.3); ladder-towers (13.2.4); demountable scaffold systems (13.2.5); hydraulic extending lift platforms (13.2.6); and trailers and vehicles with hydraulic extending lift platforms (13.2.7).

**13.2.2** Step-ladders, small enough to be carried by one man, will give up to about 2 m to the top step, permitting reach up to about 3.5 m, or a little higher if a hand-steady is provided at the top. It is not usually possible for a man to reach both ends of a 2 m fluorescent luminaire conveniently from one position.

**13.2.3** Fixed-length and extending ladders may be used for access to heights of up to about 9 m. Many industrial injuries arise from the use of such ladders, which for use over any period should be fixed at the head and foot. Where such a ladder is being used to gain access to luminaires mounted on trunking or under roof joists, the head of the ladder can be fitted with purpose-made hooks so that the ladder cannot topple sideways, but the foot must still be anchored for safety or the ladder 'footed' by an assistant. A safety-harness is available for use when working at heights of 2.5 m or more above the ground.

**13.2.4** Ladder-towers, constructed of wood or metal, consist of nesting frames which can be scaled up to heights of around 7 m. The tower is usually provided with a base having wheels for movement, and extendable *outriggers* to stabilize the unit when extended. The maker's instructions for use should be followed in respect of locking the wheels against movement or screwing down the feet on the outriggers before scaling, and it is important to use such units only on firm ground. When fully extended, some tower-ladders are unstable if subjected to even moderate wind-forces and must be stayed out of doors or in windy conditions.

**13.2.5** Demountable scaffold systems consist of frames which are placed upon one another to achieve the required height. There may be built-in ladders positioned internally, and platforms formed at the head and at intermediate heights. The provision of hand-rails is usually part of the system. Follow the manufacturers' instructions regarding the maximum number of units that may be assembled vertically. A typical maximum platform height for an un-stayed assembly is around 12 m. These units must only be used on firm level ground, or should be erected on firm planking. One system provides wheels for the lowest scaffold frame, permitting the unit to be moved while erected, and the wheels must be locked, or the levelling feet screwed down to prevent movement. A variation of the principle used in this sort of equipment is the assembly of a mobile 'bridge', which provides a means of access to luminaires situated over a floor obstruction; thus, the two wheeled sections can stand in adjacent aisles to give access to luminaires over a row of machines.

**13.2.6** Hydraulic extending lift platforms are used for heights up to about 20 m, the cradle providing accommodation for two men. Extension may be achieved by hand-pumping or by an electric pump connected to a nearby mains-socket or to rechargeable batteries in the base of the unit. Control of height may be effected from the cradle, and the cradle may have an offset to enable things to be reached 2 m or more off the centre of the unit. A feature of the construction may be the facility to fold in the wheeled feet to permit the unit to pass through a doorway or narrow space. It is essential that these pivoted feet are locked in the operating position before scaling.

**13.2.7** For use on major installations with the need for high access, there are trailer-towers and tower-vehicles giving access up to around 15 m, and far higher than this for special applications. Such sophisticated and costly access plant is rarely needed for interior lighting installations, for it is generally more convenient and cheaper to provide some form of *permanently-installed* access in the building (13.2.8).

**13.2.8** A large part of the cost of maintenance is the labour cost involved in bringing the man to the luminaire, and, where mobile access equipment is used (13.2) there is the cost of moving the equipment to the job, plus its return to the place of storage, and the cost of storing the access equipment when it is not in use. Where it is possible to build in the means of access, although the initial cost may be higher, the maintenance engineers may be able to get to the equipment with minimum delay and without the disturbance to production occasioned by having to move ladders and other equipment through the factory during working hours. Types of built-in access used in high roofed buildings (8.3) include *high-level walkways* (13.2.9), *through-the-roof access* (13.2.10) and *access from the gantry crane* (13.2.11).

**13.2.9** Access to high-mounted luminaires can be gained *via* walkways constructed within the roof framing, the walkways providing a safe means of access even at great heights. The walkways should be wide enough for two men to pass each other (say 750 mm wide), and be provided with a waist-high hand-rail on each side (1 m high). At the edges of the walkway there should be a kick-board, say 150 mm high, both for personal safety and to prevent objects slipping off the walkway and falling. The engineer on the walkway should be able to reach the luminaires without placing himself in a precarious situation, and preferably the luminaires will be pivoted so they may be brought into a safe position for attention (Figure 13.1).

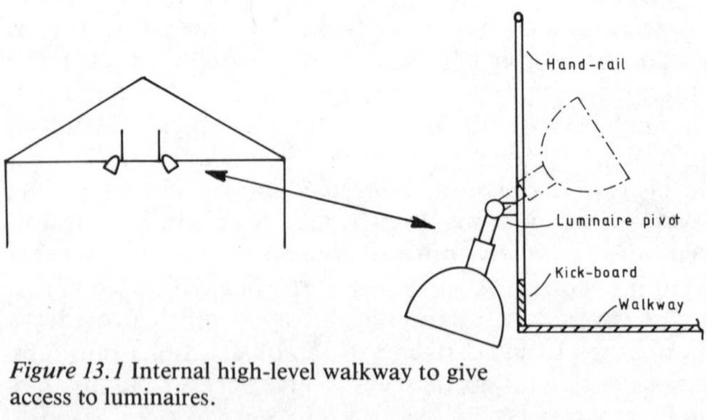


*Figure 13.1* Internal high-level walkway to give access to luminaires.

**13.2.10** There are some situations in regard to buildings which are hazardous zones (12.1) where it is necessary to place the luminaires above the roof in the normal atmosphere, and to direct the light through roof windows into the interior (Figure 12.2), as may be required where the hazard consists of hydrogen gas for example. The idea of using roof-access can be used in normal atmosphere buildings, though it may be preferred to mount the luminaires inside the roof. Then, access may be gained by the use of hinged panels in the roof, and in some cases the luminaires will be brought outside for attention.

Again, the walkway should be constructed to ensure a good standard of safety for the maintenance engineer, and it is recommended that the construction and dimensions given for interior roof walkways (13.2.9) be followed.

**13.2.11** Where there is an internal gantry crane, it can sometimes be arranged that this can be used to give access to the luminaires. It is important that there should be proper provision for the safety of the personnel, and a safe platform provided with handrails and kick-boards as specified for internal roof walkways (13.2.9). Particular attention should be paid to the proper guarding of mechanisms on the crane, and guarding to prevent personnel coming into contact with the electrical conductors supplying the crane drives. Ladderways should be safety-hooped from 2.5 m above the ground in accordance with British Standard practice[53].

**13.2.12** Another practical and widely-used form of built-in means of access to high-mounted luminaires is the use of *raising-and-lowering gear*. This employs a small winch for each luminaire, by which means each may be lowered to the ground for cleaning and relamping. Above the luminaire is a connecting unit which carries the weight of the luminaire in normal operation. To lower a luminaire, the winding-handle is inserted in the winch, and turned as to raise the luminaire, an action which disengages the electrical connection to the luminaire and releases the physical support. Then, turning the handle in the opposite direction will lower the luminaire. To return the luminaire to service, it is wound up to position; further pressure on the winding-handle engages the mechanical lock in the connecting unit, and restores the electrical connection also. Thus, in normal service, the weight of the luminaire is not carried by the cable (Figure 13.2). Precautions against mishap in the use of this equipment should include ensuring that any gantry crane is not operated while luminaires are being lowered. It should be noted that, although good quality raising-and-lowering gear as described is usually highly reliable, should there be a mechanical or electrical fault at high level, some means of gaining access to the cable pulleys or connecting units over the luminaires will be needed.

## 13.3 Preventative maintenance and relamping

**13.3.1** The technical life of an interior lighting installation in industrial premises is often in excess of 20-years, though it is customary to amortize the capital cost over a period of 10-years for taxation purposes, and to take that period as the basis for economic comparisons (18.2). The purpose of preventative maintenance is to keep the lighting installation in good condition, so that it gives the lighting service that is needed from it, and to keep the system safe. Lighting installations that are not cleaned and inspected periodically will give lower lighting output, and over a period of time may produce risks of electric shock or fire due to neglect to repair damage due to mishaps and deterioration. Preventative maintenance will include cleaning of

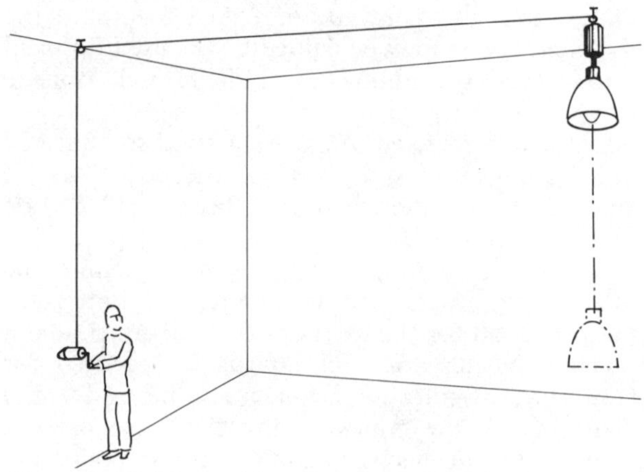

*Figure 13.2* Raising-and-lowering gear for high bay luminaires.

luminaires (13.3.2); steps to reduce rust and corrosion (13.3.3); electrical testing (13.3.4); and the periodic replacement of lamps (13.3.5).

**13.3.2** The cleansing procedure for luminaires will depend upon the type of luminaire and the degree and nature of the atmospheric pollution at the place of use. Conventional types of luminaires for fluorescent tubular lamps are usually cleaned *in situ* by wiping with a soft cloth wrung out in hot soapy water and then wiping with a clean cloth. The luminaire should be switched off before this work is done, and the lamp should be removed. Enclosures and diffusers are usually brought to ground level for washing in a deep vessel, and may be dipped in a solution of an 'anti-static' material (which reduces attraction of dust particles and helps keep the plastic components cleaner). It is time-saving to have some spare enclosures, and to replace a batch at a time. High-bay luminaires may be cleaned *in situ*, unless they are provided with 'over-lamp' reflectors, or if the luminaire is detachable for cleaning purposes. The cleaning operation timing should be adjusted if necessary to make it coincide with time for a relamping (13.4.5). The cleaning operation should be timed to fit the planned period which has been estimated as yielding the Maintenance Factor used in the lighting design calculations (16.2). The correctness of this timing can be judged by periodic measurement of the illuminance at selected test positions. No abrasive or gritty materials should be used for cleaning luminaires, nor caustic materials which may damage the paint finishes. The work of cleaning should be combined with steps taken to reduce rust and corrosion (13.3.3). Even luminaires with 'restricted breathing' (e.g. dust-tight) will aspirate dirt and dust into their interiors over a period of time, and will need to be opened and cleaned. The cleaning of 'proof' luminaires (e.g. Flameproof, for Hazard Zones) should be carried out only by responsible trained persons, for the correct re-assembly and proper tightening of all fixings is vital. This work can only be done when there is no flame hazard and a permit-to-work has been issued (Chapter 12).

**13.3.3** In dry 'ordinary atmospheres' luminaires should not normally be troubled with rust or corrosion, but even a short period with the building unheated may produce signs of deterioration. This can be minimised by lightly spraying all exposed metalwork with a silicone rust-inhibiting spray after cleaning. Hinges of openable metal parts may be similarly sprayed or very lightly greased with high-melting-point grease or petroleum jelly. To prevent jamming of lamps in E.S. and G.E.S. lampholders, the threaded portion of a new lamp may be given an *almost imperceptible coating* of graphite grease—but care must be taken not to allow this to get to other parts of the lampholder as it is conductive and will cause short-circuiting. Where greases or sprays are used on luminaires, avoid contaminating gaskets or elastomer sealing components with any substance incompatible with the gasket or seal material which could cause its deterioration. During the maintenance procedure, the operative should examine all cable runs, conduits and fittings for damage or corrosion, and take appropriate steps if damage or deterioration is found.

In high buildings where entry into the roof area is infrequent, the inspection should include signs of condensation or roof leakage which could affect the lighting equipment. Condensation from steam (e.g. in the paper and food industries) will form on cold roof structures during the winter, and the amount of moisture may be deleterious to the luminaires and electrical installation.

**13.3.4** Periodic insulation testing should be carried out according to the requirements of the *Wiring Regulations*(15). Note that electronic components in transistorized control-gear and ignitors used with some HID lamps will be damaged if subjected to over-voltage or reverse-voltage as may be applied during an insulation test; therefore such circuits should be shorted-out during insulation tests.

**13.3.5** Until very recently, it would undoubtedly have been excellent practice to carry out the bulk-replacement of all lamps at the approach of the expiry of their nominal lives(40), but, with the advent of various developments in lamps, the matter is not as clear-cut as previously. Lamp replacement procedures are best discussed under headings of the types of lamps:

*Tungsten-filament lamps:* (GLS)

These give a substantially constant light output through life, the light output falling off a few per cent near end-of-life. Failure is total, and virtually without warning. In any batch of lamps, there will be sporadic failures through life, but the average life for a batch will be 1000 h for standard lamps, or 2000 h for 'extended life' or 'double life' lamps. Replacement near the end of life, or say when 5 per cent failures has occurred is usual.

*Tungsten-halogen lamps:* (T-H)

Life characteristics are generally as for tungsten-filament lamps, but the nominal life is 2000 h. Replace at 2000 h or when 5 per cent of the lamps have failed.

*Mercury-vapour lamps:* (MBF, MBFR)

These have a nominal life of 7500 h, but operating under conditions of normal voltage and temperature and free from excessive vibration they may remain in operation for 12 000 to 14 000 h. Their light output declines through life, with a typical depreciation characteristic of giving their *lighting design lumens* at 2000 h, and about 0.8 of their *initial lumens* at 7500 h. Beyond this time the depreciation continues to around 0.65 of the initial lumens at 12 000 h. It is thus generally economic to replace this type of lamp at around 7500 h, and certainly before 10 000 h. (In roadlighting practise it is customary to run the lamps to failure, but this is neither convenient, economic nor safe in industrial lighting installations.) Beyond 7500 h there may be a slight shift towards a blue-er colour-appearance, with some loss of colour-rendering property. Bulk-replacement at 7500 h is recommended.

*Metal-halide, mercury-halide lamps:* (MBI, MBIF)

When first marketed, some makers offered these lamps with only 5000 h life, but experience has shown that they can give effective lives greatly in excess of this. The latest formulations of these lamps have not been in use long enough for reliable ultimate-life figures to be ascertained, but it should be noted that the characteristics of these lamps from different lamp-makers may differ significantly in colour-appearance, colour-rendering, life performance and electrical characteristics. In general, these lamps have a colour-shift towards a pink-er colour-appearance through life, though the colour-rendering is fairly constant. The lumen-maintenance is not too good, with lamps giving outputs typically 0.7 of their *initial lumens* at 7500 h; because of this, the lamp may be offered with a nominal life of only 6000 h. This is one reason for the declining popularity of these lamps, which do not appear to have lived up to their makers' early expectations. As individual lamps which have been replaced may appear to be markedly blue-er and brighter than the other lamps in the installation, in situations where appearance matters it will be best to replace this type of lamp at not later than 6000 h.

*Fluorescent tubular lamps:* (MCFU)

These lamps in powers of 20 W and over have a nominal life of 7500 h, but, burning continuously, or with fairly infrequent switching, under conditions of normal voltage and temperature, they may give a life of up to twice this. However, there is a marked depreciation of lumen output through life; as near as can be judged an output of 0.8 of the *lighting design lumens* is reached at end of nominal life, and the output can be as low as 0.5 of the l.d. lumens at 15 000 h. It is therefore difficult to justify the use of these lamps much beyond nominal life, and bulk-replacement at 8000 h or two-yearly intervals is recommended. About the end of life, there may be some difficulty in starting these lamps, a difficulty that increases as the lamps age further, and particularly noticeable in low ambient temperatures. Starter-cannisters (other than electronic starters) for switch-start lamps should be replaced not less frequently than at every other relamping. Better-colour-rendering tubes used for colour-matching work (4.2) should be replaced at 3000 h for the finest work, and at 5000 for ordinary industrial colour work. Because of the colour-shift that occurs with age, it is not usually practicable to replace individual lamps in a set, e.g. in a film-viewing frame, and it is customary to operate some additional lamps outside the apparatus, but switched in parallel with it. Thus, these 'spare' lamps will have expended the same number of hours of operation as those within the apparatus, and can be inserted to replace any that fail before the time for bulk-replacement.

*High-pressure sodium-vapour lamps:* (SON series)

The life claims for these lamps are not at all clear, and some patterns and powers, at the time of writing, have not been in use long enough for the situation to be defined. Initially, they were offered tentatively at 6000, then at7500 h nominal life, but the smaller powers in roadlighting installations give lives of as much as 20 000 h, and it may well be that possible life may be as much as 30 000 h. Attractive though this is, there is a snag, in that the lumen maintenance is only fair, though this is offset by the characteristic of sodium lamps to draw more energy from the mains as they age. The nett lumen output per lamp remains fairly constant, but the true efficacy (lumens per watt) declines progressively as the lamps age. Where the cost or difficulty of replacement is great, it may be judged economic to run these lamps for periods of 10 000 to 20 000 h, but this is at a steadily increasing energy cost. (In public lighting, as the cost of replacement is high, some Local Authorities are running these lamps to failure. The extra current drain by this unmetered load will doubtless have to be paid for by increased charges by the Electricity Boards to Local Authorities in the future. A somewhat similar situation exists with regard to *low-pressure sodium-vapour lamps* (SLI, SOX), which, being unsuitable for industrial use, are not dealt with in this book.)

**13.3.6** The *maintenance factor* (16.2) used in calculation of the lighting system may be confirmed or adjusted by taking lightmeter readings, and calculating the actual Light Loss Factor (16.2.5) of the installation, using the formulae given in the *CIBS/IES Code for Interior Lighting*[5]. By making these measurements the maintenance engineer may choose lamp replacement periods and cleaning periods which will yield the lowest possible cost-in-use (18.2).

**13.3.7** The light output from lamps is not uniform over time, but generally tends to decrease (though, in the case of SON lamps, the declining efficacy through life tends to be balanced by an increase in current drawn from the mains). The light emitted by luminaires is not constant either, but tends to decline due to the accretion of dust and dirt. On being cleaned, the light output is restored almost to the maximum level, but there is always a small percentage of permanent increased loss due to irreversible staining or corroding of reflecting surfaces or the darkening of transmitting media (e.g. the *solarization* or yellowing of certain plastics materials caused by exposure to ultraviolet raditation). The nett light output tends to follow a saw-tooth pattern through the life of the installation (Figure 13.3), and it is apparent that a significant saving in energy may be made by proper attention to regular cleaning of lamps and luminaires[5][40].

## 13.4 Compatibility of discharge-lamps with control gear

**13.4.1** Modern fluorescent lamps and discharge-lamps last a long time in service. One UK lamp manufacturer gives a written 8000 h pro rata guarantee for their SON lamps, and other manufacturers may also compensate a user if the lamps fail to produce the average life (for a group of lamps) that is claimed by the maker. It is possible for lamps to be in the socket for as long as six years or even longer when the hours of use are fairly short; after such a long period, the person responsible now for relamping may have no knowledge of the original lamp specification, and possibly cannot find a written record. Thus, if he has a stock of similar, but not identical, lamps, or if the original type or brand of lamp is not available, or possibly that an alternative make is now offered at a lower price, he may be tempted to gamble on the alternative lamp being suitable to use with the existing control-gear. It is therefore recommended that a schedule of correct lamp types for all

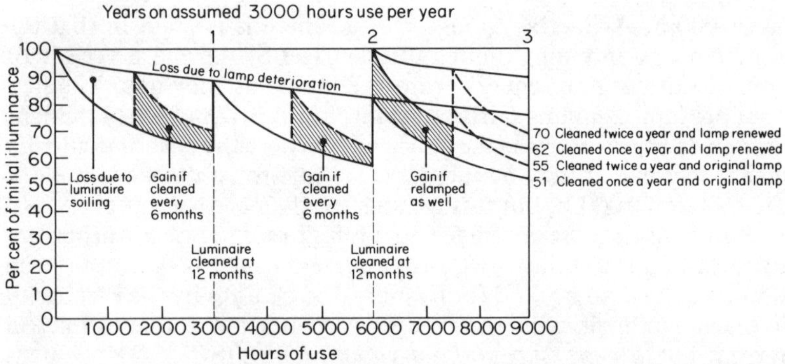


*Figure 13.3* Depreciation of illuminance. The diagram assumes 3000h of use per annum for a fluorescent tube or HID lamp installation, and shows the characteristic saw-tooth curve of illuminance over time due to lamp light loss and the light loss due to soiling of the luminaire and room surfaces. It shows that more frequent cleaning of the luminaires enables a desired average or minimum illuminance to be achieved with a lower initial light output, and thus with a potential saving in installation cost and cost of energy. But the cost of the extra cleaning must be set against such savings. The diagram is typical only; a forecast of the end of lamp life minimum illuminance or of the service illuminance can be calculated by applying a *light loss factor* (16.2) as described in the CIBS/IES Code.[5].

luminaires in an installation is compiled, and kept for reference; a copy could be placed in a sealed plastic bag and kept near the switchgear. But even this precaution will not solve the problems regarding new kinds of alternative lamps which come on the market long after the luminaires are installed. If the lamp used is not compatible with the control gear, then there may be difficulties in starting, short lamp life, excessive and costly current drain, in some cases damage to the luminaire or control gear due to excessive temperature rise which sometimes can be a cause of fire, or the lamps may be unstable in light output or colour.

**13.4.2** Except for certain 'plug-in' and 'energy-saving' alternative lamps for existing luminaires that are offered by reputable manufacturers, it is sound policy to always relamp installations with the same make and exactly the same type of lamp as was originally specified for the luminaires unless a check is made with the lampmakers and the luminaire manufacturers. The lighting industry is sensitive on this subject of the compatibility of one maker's lamp on another maker's control-gear; any recommendations from the original supplier to use only his lamps sounds very much like a selling ploy, but the advice may be sound. Incompatibility of lamp to gear may arise because of (a) differences in the ignition or starting means, (b) mis-match between the electrical characteristics of the lamp and control-gear, or (c) non-interchangeability of lamp-caps between nearly similar lamps. It will be assumed that no responsible person would attempt to use a different lamp shape (e.g. tubular bulb, eliptical bulb, reflector bulb) to that for which the luminaire was designed.

**13.4.3** It can happen that the maintenance engineer or the electrical contractor goes to purchase replacement lamps from a wholesaler, and is offered an alternative brand or type of lamp. When the user or the electrical contractor fits lamps other than those approved by the luminaire-maker, he probably relieves the makers of the lamps and the control-gear of the legal responsibility for any damage or mishap which may result, and takes this responsibility upon himself. The product supplier is liable to the user for loss or damage arising from the use of his product (and probably from its mis-use, if not warned against). Lighting manufacturers print general warnings against mis-matching lamps and gear, and state specific restraints in their product literature and instruction leaflets. Because of the extension of lamp lives, and the possibility of using new kinds of lamps, the user or contractor may want to know if an alternative lamp may be used. The counsel of perfection is 'Don't—but, if you must, check first with the makers of the lamp and the control-gear'. Unfortunately, the manufacturers cannot always give unequivocal advice, simply because they have not tested every combination of their lamps and control-gear sets against all those available from other manufacturers, there being hundreds of possible combinations. For this reason it is impossible to publish a definitive table of 'fits and mis-fits'. In a paper to the IES/CIBS National Lighting Conference, 1978, Keward, Ogden and Parker[46] reviewed the performance of discharge-lamps on 'alien' control-gear. Extracts from their data, updated (1980) with information from some manufacturers are the basis for the guide lines given in this Section.

**13.4.4** It is emphasized that the fact that a lamp lights is not a positive indication that all is well. Because of manufacturing tolerances, within any installation some lamp/luminaire combinations may be satisfactory and others not. True compatibility is only established after extensive experience in actual installations, with careful monitoring of the failure-rate of lamps and luminaires. Certain combinations may give trouble only after several thousand hours operation; some may mis-operate when the line voltage is below the nominal rating; others may give trouble in high or low ambient temperatures.

**13.4.5** The long-established BS fluorescent tubular lamp ratings (20, 40, 65, 80 and 125 W) do not present problems of compatibility, and existing luminaires will usually accept standard tubes of other makers if of the same nominal rating and diameter. However, two lamps of the same general description but obtained from different makers will not necessarily be of identical colour-appearance (4.2.2) and in some cases will not produce the same colour-rendering (4.2.5). For these reasons, mixtures of lamps from different makers in the same installation are not recommended.

**13.4.6** There are now available 'energy-saving' tubes, designed to be used in existing standard luminaires which, by virtue of a krypton gas additive, consume 10 per cent less energy than the standard tubes they replace, but with little or no loss of lumen output. Such tubes, (for example, the Philips TLD and the Thorn 'Pluslux' range) are of 26 mm (1 inch) diameter, and come in 18 W (600 mm), 36 W (1200 mm), and 58 W (1500 mm) ratings. Also available in the same ratings are the multi-coated or polyphosphor tubes, (for example the Thorn 'Polylux' tubes and the Philips TL/80 tubes) designed to give improved colour-rendering and offered as alternative to standard 'Natural' tubes. Philips offer their 'Powerslimmer' luminaires with their TLD lamps; for example, a luminaire with an acrylic controller (suitable for use in offices and non-manufacturing areas) which takes two 58 W, 26 mm (1 inch) diameter tubes, and such luminaires should not be relamped with standard 38 mm (1.5 inch) tubes.

**13.4.7** Krypton-dosed 2400 mm T12 tubes rated at 100 W (for example the Thorn POP-100 tubes) are compatible with existing 125 W loading *switchstart* control-gear, and give about 10 per cent less lumen output than standard 125 W tubes. Also, special 100 W luminaires are available for these tubes, but these should not be relamped with standard 125 W tubes.
**13.4.8** Wotan† produce 65 W tubes (L65W/..UK and L65W/..UK In) for use on their special ballast which operates at a lower lamp voltage and a higher lamp current than is used in standard 65 W switchstart circuits. This Wotan circuit is not compatible with standard 65 W tubes, and the Wotan tubes mentioned may not achieve their designed performance in other circuits.
**13.4.9** Turning now to high-intensity discharge (HID) lamps, it should be noted that the lamp makers in the UK have not standardized their designs for all these lamps, and in some cases there are important differences between the lamps of various makers. For example, there are three non-interchangeable cap arrangements for double-ended tubular SON lamps on the British market: Thorn 250 and 400 W ratings use caps similar to those on linear tungsten-halogen lamps; Wotan use Fc2 caps; Osram (GEC) use R12.5 caps. Apart from these, all SON lamps of equivalent type and rating in the UK market are interchangeable. Lamps which are physically interchangeable are not necessarily suitable alternatives; for example, the 'high arc voltage' and 'low arc voltage' versions of the 1kW MBF lamp look similar, but have quite different electrical characteristics. While progress towards standardization is welcome, it is through individual lamp-makers striving for commercial advantage by developing new products that has resulted in the recent impressive developments in lighting technology. The best-designed, or perhaps the best-sold or cheapest product outsells, and eventually outlives the others, and may in time become the standard. It may be expected that there will be considerable changes in lamp technology in the 1980s; we are going through a time of innovation and development, stimulated by the need for energy-conservation and economy.
**13.4.10** Because users are reluctant to scrap existing installations, there is interest in 'plug-in' replacements to bring the advantage of improved efficacy without the immediate necessity of replacing the luminaires. Even today, there are still tungsten-filament lamp installations which, at least in the short-term, are worth re-lamping with tungsten-mercury (MBTF) blended lamps, at no cost other than for replacing the lamps in their sockets. Alternatively, luminaires designed for use with tungsten-filament lamps can be locally rewired for conversion to MBF lamps. One way of doing this is to use pre-wired control-gear boxes (such as the Osram (GEC) 'Wattsaver' range) mounted local to the luminaires, enabling the replacement of t.f. lamps of 100, 150 and 200 W ratings with 80 W MBF lamps, and those of 150, 200 and 300 W ratings with 125 W MBF lamps. Such conversions make effective energy reductions, combined with significant improvements in lighting levels.
**13.4.11** In general, the long-established BS ratings for MBF lamps present

†Wotan is the trade name in the UK of Osram GmbH, which has no connection with the British company of Osram (GEC) Ltd.

no problems of compatibility between lamps of the same rating from alternative manufacturers. This is not the situation in the case of Metal-Halide (MBI) lamps, where, because each lamp manufacturer uses different arc-tube additives and pressures, all the MBI lamps on sale in the UK are compatible only with the control-gears supplied by the lamp-makers. Attempts to re-lamp MBI installations with lamps of another brand will almost certainly give rise to starting difficulties, to short lamp life, to damage to control-gear, or to unstable light output or colour.

**13.4.12** The situation in regard to high-pressure sodium (SON) lamps is complex, in that the method of providing the electrical starting impulse differs between types and makes. For standard SON ratings of 150 W to 1000 W, the lamps made by Crompton, Iris, Philips, Thorn, Sylvania and Wotan are generally compatible with all these makers' control-gears. But, single-ended SON lamps in this range from Osram (GEC) are fitted with internal 'snap-starters', and do not need external ignitors (though their 'Solarstream' SON-L lamps, which do not have an internal starting device, do). Existing installations of SON lamps fitted with internal starters can only be re-lamped with lamps of the same type, unless external ignitors are wired in at some additional cost. 50 W and 70 W SON lamps may be fitted with internal glow-starters, though many existing 70 W lamps are operated by external starter-switches. Modern ignitor devices for SON lamps give almost instant starting from cold, with a hot re-strike time of less than a minute. Lamps with internal starters may take up to 10 minutes to hot re-strike after an interruption of supply. It would be good practice in installations fitted with SON lamps having internal starters, to provide a proportion of the luminaires with external starters of some kind, to ensure rapid restoration of lighting after any power interruption.

**13.4.13** There are two ranges of SON lamps, (a) conventional SON lamps to operate on SON control-gear, and (b) 'plug-in' lamps to replace existing MBF lamps and to operate on MBF control-gear. The latter type must have an internal starting device, either a mechanism (e.g. a snap-starter or a glow-switch) or, as in the Philips' lamps, an auxiliary electrode; some rely on a modified gas-filling to enable the lamps to start without a high-voltage impulse, but this involves a loss of up to 30 per cent of the lumen output as compared with lamps having other starting methods. Plug-in lamps (apart from the Osram (GEC) 310 W rated lamp, which requires a tapped choke) operate at higher currents than the original MBF ratings; this over-running of the gear can result in over-heating, particularly in the case of MBF gears of old design which do not comply with present IEC standards. The repeated attempts of a lamp to strike by the action of an internal starting switch could possibly cause damage to control-gear, or could cause cable breakdown; thus, lamps with internal starters should only be used with very good quality control-gear, in approved ciruits, and with 450/750 V grade cable between the lamp and the control-gear.

**13.4.14** When a manufacturer offers a lamp or tube for use in luminaires of his own or another makers' design, and states in his advertisements and leaflets that it is a suitable replacement for the lamp or tube originally specified, there can be no doubt that the lamp-maker is confident of the compatibility, and accepts responsibility for successful operation. It is unwise to experiment with combinations not specifically recommended or approved by the makers. The foregoing resumé of the situation cannot be claimed to be complete, though efforts have been made to ensure accurate cover of the essentials by consultations with some of the leading lamp-makers. The situation will change with the arrival of new products on the market. Whilst lamp and luminaire makers are reluctant to publish general guidance on the compatibility of their lamps and control-gears with those of other makers, it is understood that they will advise on specific proposed combinations where it has been possible for them to carry out tests and to obtain reports of field trials.

Chapter 14

# Portable and mobile lighting in the factory

Apart from use under emergency conditions (e.g. during a mains-failure) (Chapter 6), portable and mobile lighting can serve a useful role in the factory (14.1). Such lighting may be needed from time to time for maintenance and construction work, and for occasional activities for which the cost of permanently-installed lighting is not justifiable (14.2). The important thing is to avoid dangerous improvisations, but to have ready the means for providing lighting for those occasions when the normal lighting will not be adequate or suitable. For repairs, machine-tool setting, inspection inside metal chambers and in 'earthy' locations, reduced-voltage systems may be needed (14.3); consideration might be given to installing a reduced-voltage distribution system (14.4) to avoid entirely the use of mains-voltage portable lighting in the factory, and thus provide a higher standard of electrical safety.

## 14.1 Hand-lamps, battery-lamps, trolley-lights

**14.1.1** Battery-powered hand-lamps have only a limited use in factories, and there are problems of shelf-life and replacement of batteries. Pilfering of such torches is common and almost impossible to prevent. Substantially-sized hand-lamps with re-chargeable batteries are far more practical, and can be left plugged into the charger when not in use. Portable self-powered units such as these can be useful and safe for use in enclosed spaces such as ducts, or inspecting in confined areas, such as in the base compartments of machine-tools.

**14.1.2** A useful device is a *trolley-light,* which may consist of a small-wheeled trolley carrying a secondary-cell battery, complete with its charging circuit. On the trolley may be mounted a 12 V spotlight on an adjustable arm, and one or more 12 V hand-lamps may be plugged in as required. For safety, such a unit should be unplugged from the mains when in use, and it must be a strict rule that trolley-lights suitable only for use in normal atmospheres are not taken into hazard zones (12.1).

## 14.2 Engine-driven mobile stand-by sets

**14.2.1** This type of equipment consists of a small-power petrol engine driving an alternator, the power unit mounted within a carrying frame. When fitted with a collapsible mast and light, these units are commonly known as 'Jenny-lights', and are widely used for temporary exterior lighting[1]. They can only be used indoors for short periods in very well ventilated areas because of the exhaust fumes, but are sometimes fitted with a flexible metallic hose extension to the exhaust pipe to enable them to be vented out a window or other opening. As they are usually suitable for outdoor use, they may be positioned outside and a cable brought inside for temporary use, when it is important to ensure that the Jenny-light circuit is properly connected to the electrical 'earth' ('ground') of the building. A Jenny-light may be mounted on wheels for hand movement over fairly smooth ground (Figure 14.1), while there are available various size mobile units ('trailer-lights) giving power outputs of up to 7.5 kW. These units can be positioned outside a building (to avoid nuisance due to exhaust fumes and engine noise) to provide a temporary supply within the building.

**14.2.2** Engine-driven mobile stand-by sets are of value in taking over the duty of the emergency lighting system when ever the latter is out of commission for repairs or for periodic discharge/recharge of the batteries (6.1.7).

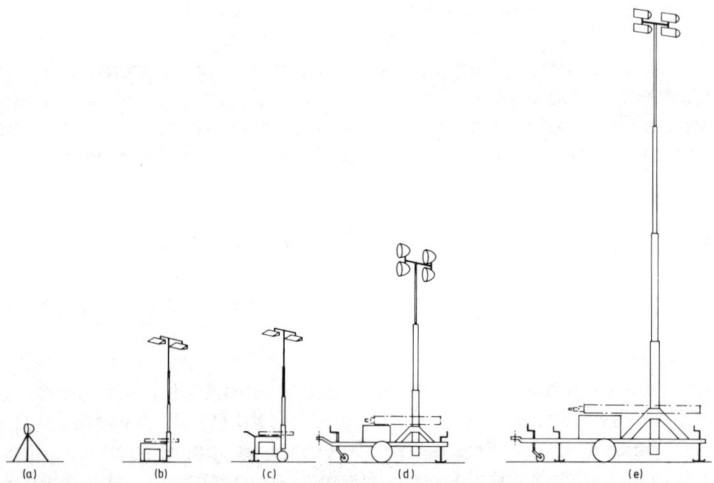


*Figure 14.1* Exterior self-powered lighting equipment. (a) Tripod-lamp for operation from one of the powered units. (b) Jenny-light, and (c) Wheeled Jenny-light, typically of between 350 W and 1500 W power output. (d) Medium-sized trailer-light, typically of around 3.5 to 4.5 kW power output. (e) Larger trailer-light, typically of 4.5 to 7.0 kW power output.

## 14.3 Reduced-voltage portable lighting

**14.3.1** General lighting systems provide a spread of light over the whole working area; localized lighting provides enhanced illuminance at work-stations; local lighting under the control of the operator gives the facility to direct light at high intensity on small areas plus the ability to direct light at angles beyond the reach of overhead lighting (2.1, 2.2). *Portable lighting* is for occasional use, for example during breakdown repairs or periodic maintenance procedures, or during the erection of new plant. Experience shows that portable mains-voltage portable lighting has inherent risks of electric shock, and, although there is no such thing as a 'safe voltage', operation at *reduced-voltage* greatly minimizes the dangers (14.3.2).

**14.3.2** The reduced-voltage commonly used in industrial applications and on building and construction sites is 110 V in the UK, though there is a possibility that this will ultimately be standardized at 100 V to agree with EEC practice. The reduced-voltage is derived from a double-wound transformer having an earthed screen between the primary and secondary windings. To reduce the risks further, the maximum potential between line and earth is halved by centre-tapping the secondary winding solidly to earth (Figure 14.2) giving a maximum potential between appliance or line and earth of 55 V a.c. under fault conditions. The transformer is enclosed in an earthed metallic enclosure, or may be enclosed in an all-insulated box. The enclosure may be drip-proof for indoor use, or fully protected for outdoor use.

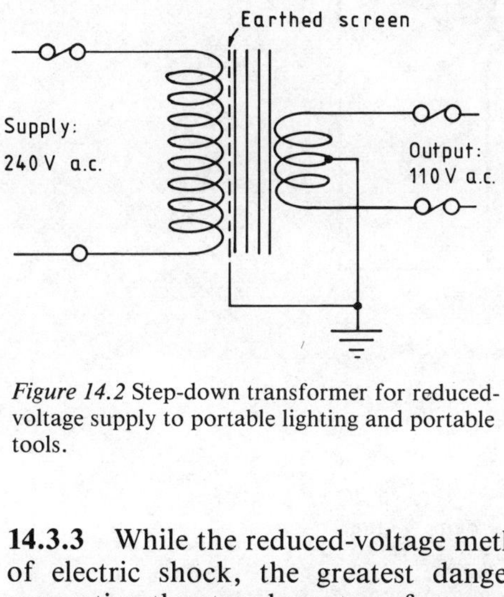


*Figure 14.2* Step-down transformer for reduced-voltage supply to portable lighting and portable tools.

**14.3.3** While the reduced-voltage method will certainly reduce the hazard of electric shock, the greatest danger remains the mains-voltage lead connecting the step-down transformer to the supply. Therefore, this mains lead should be as short as is convenient, and preferably the cable will be protected in a flexible metallic hose. The output cable from the transformer to the appliance may be pvc/pvc flexible, the external sheath being yellow to signify reduced voltage. The preferred method of connection is by a BS 4343:1968 industrial plug[41] into a matching socket on the transformer case. (These plugs, sockets and couplers are colour-coded and pin-orientated for voltage to prevent misconnection). The use of this equipment is certainly preferred to improvisations using portable lighting at mains voltage, but, if the need to use portable lighting in the factory is re-occurring, it would be better to equip with premises with a permanent reduced-voltage distribution system (14.4) to provide the supply to all portable lighting and any lighting equipment that is installed within normal hand-reach height from working positions.

## 14.4 Reduced-voltage distribution systems

**14.4.1** Where the need for portable lighting is re-occurring, for example, in the paper industry, where it may be necessary to carry a light inside a papermaking machine while lacing-up or clearing jams, it would be preferable to install a permanent system of reduced voltage distribution, with socket-outlets placed at all positions where the need for portable lighting is likely to be needed. An excellent precedent for this is modern practice on building and construction sites[42] where distribution systems to BS 4363[43] are employed. The safe principles of that Standard are readily applied to permanent installations. The socket-outlets should be to BS 4343[41] (14.3.3), to prevent any possibility of a voltage mis-match. In large premises, there could be several step-down transformer cubicles, to obviate the need for long cable runs at the reduced voltage (Figure 14.3).

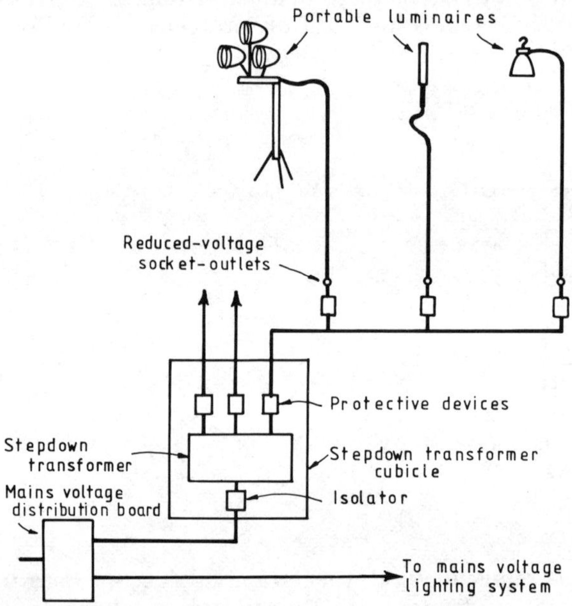


*Figure 14.3* Reduced-voltage distribution system for a permanent factory installation.

Chapter 15

# Exterior lighting

While the present volume deals mainly with lighting for industrial interiors, the lighting of the exterior spaces is also of importance and should be correctly related to the interior lighting. Technology and practice in exterior lighting has advanced considerably in recent years[1], and it is better appreciated how lighting of factory open areas can contribute to efficiency and safety (15.1), and how good roadlighting on factory roads facilitates vehicle movements (15.2). An entirely new technology and philosophy relating to security lighting has grown up in the last decade, and is briefly summarized in this chapter (15.3). The amenity and prestige value of outdoor lighting at factory locations is reviewed (15.4).

## 15.1 Yards, lorry-parks, loading-bays

**15.1.1** For efficient working and safety, the lighting system for any factory will extend to the illumination of the open spaces between and around the buildings, with particular attention paid to producing conditions of good visibility for vehicle movements. The lighting plan should take account of the needs of a tired driver of a heavy vehicle, arriving in the dark, who needs to identify the premises and to find a place to park safely while making enquiries. The lighting must aid the work of the security guards who will inspect the vehicle and pass it to the weighbridge (15.3). It is impossible to separate the functions of efficiency, safety and security in such lighting; it may be noted that in well-lighted lorry parks the drivers tend to park closer. Some extra 10 to 15 per cent of vehicles may occupy the space if the visibility is good. Obstructions should be lighted, and warning lights may need to be placed on jutting-out canopies.

**15.1.2** Open spaces should not be used for parking as well as for storage of palletized or cased goods. This is a situation which may lead to accidents as well as easy theft. It is better if arriving vehicles are directed to a visitors' park or 'quarantine' area, for this prevents various kinds of frauds and thefts.

**15.1.3** Lighting under the canopies of loading-bays will facilitate safe backing-in, and additional luminaires may be positioned to throw light into the backs of open lorries to facilitate loading and unloading. At some loading-bays, fluorescent-tube luminaires in robust rain-proof housings are placed on the face of the bay platform, under the bumper-beam; these not only aid reversing-in, but they show up any gap between the platform edge and the tail of the lorry, and thus may prevent injuries caused by workers tripping or slipping into the gap (Figure 15.1).

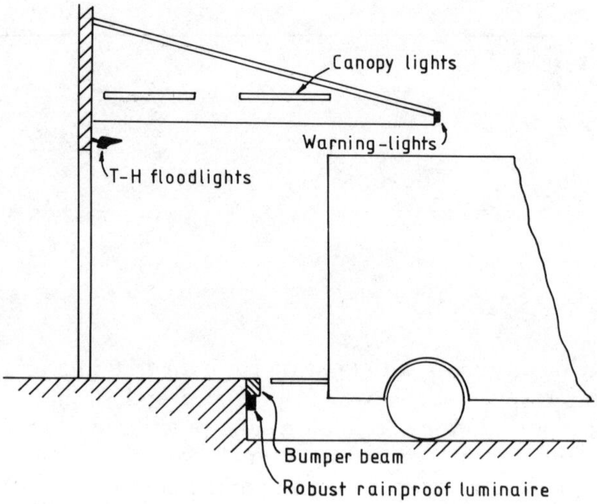


*Figure 15.1* Lighting for a loading-bay. (Not to scale.)

**15.1.4** Where workers must pass frequently between indoor and outdoor areas, thought should be given to providing a zone of intermediate illuminance if the ratio of the internal and external illuminances is high (2.3.2).

## 15.2 Factory roadlighting

**15.2.1** Only the largest of factory establishments have any considerable length of internal roadways, though many modern factory estates have common roadways which are maintained and lighted by the occupants of the factories on the estate. Generally, roadlighting to 'Group B' standards (objective luminance of 0.5 cd/m$^2$) satisfies the requirements, but for a few main estate roads the standard for 'lightly trafficked Group A roads' (1 cd/m$^2$) may be necessary. The guiding standard is BS 5489[(44)] which has replaced CP 1004. The principle is to employ *cut-off* roadlighting luminaires (beam elevation 65°) or *semi-cut-off* roadlighting luminaires (beam elevation 75°) which give rather better lighting of areas adjacent to the roadway, and which permit more economical installation design by allowing for greater spacing/mounting-height ratios. It is good practice to mount the luminaires over the kerb edge or slightly back from it to throw the riser of the kerb into shadow and thereby increase the visibility of the carriageway edge. On cambered roads, the lanterns produce T-shaped patches of luminance by reflection from the road surface, the patches meeting or overlapping to produce a more-or-less uniform road luminance. No allowance is made for the road surface condition, nor for the road being wet or dry, so the performance can be quite variable.

**15.2.2** An alternative approach which is suitable for small- and medium-sized road installations is to use post-top diffusing luminaires. These give rather more light off the road, which may be helpful for general pedestrian movement and orientation. Giving a 360° spread of light in plan, the lumens are not concentrated on the road-surface, yet these luminaires can give quite adequate illumination with an acceptable degree of glare. In tests made with a prismatic spherical luminaire (designed for 5 m mounting, but used in the experiment at 6 m) very impressive results were obtained, showing that this kind of equipment is suitable for minor factory roadlighting, for limited area lighting and for low-illuminance security lighting. Two luminaires mounted at 6 m and 50 m apart produced substantially uniform illuminance over an area measuring 50 m by 100 m, with a minimum measured illuminance ($E_h$) of 0.2 lux. Because the vertical illuminance was considerably greater, the visibility for movement and for revealing intruders were satisfactory for use in medium to low district brightness areas (Figure 15.2). The luminaires used were 'Prismaspheres' (® Holophane Europe Limited) and the full results of the trials have been reported by Baker and Lyons[45].

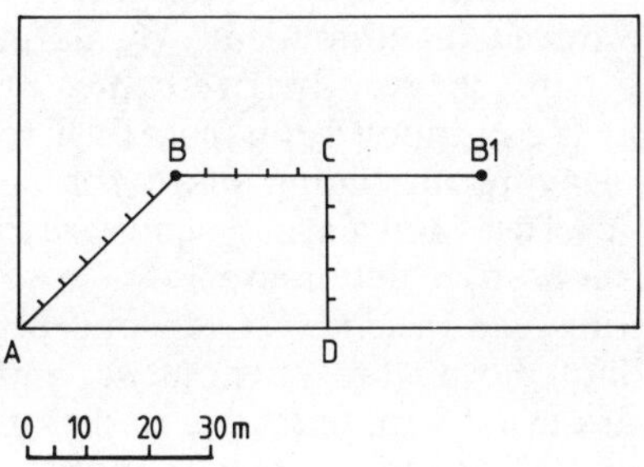


*Figure 15.2* Illuminance produced from two 125 W MBF lamp 360°-distribution post-top luminaires (prismatic spheres) at 6 m mounting. An area of 5000 $m^2$ was illuminated to 1.15 lux, with m.m.i. of appr. 0.2 lux.

| Distance along measuring line (m) | Direction B to A $E_h$ (lux) | Direction B to C $E_h$ (lux) | Direction C to D $E_h$ (lux) |
|---|---|---|---|
| 0 | 2.50 | 2.50 | |
| 5 | 2.10 | 2.00 | 0.48 |
| 10 | 1.80 | 1.70 | 0.40 |
| 15 | 1.00 | 1.70 | 0.38 |
| 20 | 1.20 | 0.54 | 0.22 |
| 25 | 0.30 | 0.50 | 0.22 |
| 30 | 0.19 | | |

## 15.3 Exterior security lighting

**15.3.1** Security lighting systems are systems of exterior lighting provided solely or mainly to enhance night security of premises, and to protect persons and property from criminal attack at night. This effect is achieved by (a) *revealing* the intruder before, during and after the attack, (b) in many cases, providing a measure of *concealment* for the defenders and thus giving them a tactical advantage, and, (c) by a combination of these two effects plus (at least in some cases) providing a lighted appearance that shows a state of preparedness in the defence of the premises, to *deter* the intending criminal.

**15.3.2** Effective systems have been developed, and are described in detail by the author in another work[1]. Systems comprise (a) *physical defences,* (fences, walls, gates etc) to slow down the attack and make it more difficult, (b) *defenders* (security guards, police, occupants etc) to observe and to act in case of an attempted entry, and (c) *security lighting*. Any form of exterior lighting has some value to security, but that which is designed for the purpose will be more effective, cheaper to install and run, and likely to be more reliable and less easy to tamper with. Sometimes lighting has an important security function, though it may appear to be only for amenity, safety, publicity or decor.

**15.3.3** Since 1970, some thousands of installations of security lighting have been brought into use in the UK, in the USA, and increasingly now in Europe.

The effectiveness of security lighting is endorsed by both military and civil security experts, by police forces and the Home Office in the UK, and by insurers. The cost-effectiveness of security lighting stems from (a) the ability to supervise long perimeters and large areas with a small defending force, (b) the physical defences need not be so stringent when suitable lighting is provided, (c) security lighting installations are of relatively low capital cost, and, as they employ low-technology, are not expensive to maintain and operate compared with more sophisticated (and often less reliable) security measures, e.g. intruder detection alarm systems. Security lighting is campatible with alarm systems, and can be very successfully integrated with closed-circuit television (cctv) systems of surveillance. However, the cctv system chosen should be one which operates with cameras requiring illuminances similar to those produced in normal security lighting, e.g. with around 10 lux on planes facing the cameras. The use of high-sensitivity (low light level cameras) for the general surveillance, and only switching on the exterior lighting when there is a suspicious circumstance is not security lighting. Such systems (which may also be switched automatically by alarm systems) are termed *trip-lighting*, and can be readily overcome by experienced criminals. True security lighting is continuously on from dusk to dawn every night throughout the year, leaving no hiding place for an intruder, his loot or his vehicle. Low-light-level cctv and alarm systems are to detect criminal activity; security lighting is a positive deterrent.

**15.3.4** The technology of security lighting uses science-based concepts involving an understanding of the dark-adapted eye, glare phenomena, and control of luminances. The installations are usually quite simple to design (often using standard arrays and layouts of lighting equipment without need to calculate) and make effective use of modern lamps and luminaires to achieve energy-economy.

**15.3.5** The techniques of security lighting are all based on traditional lighting methods, and are capable of assessment as to their effectiveness. They are (a) perimeter lighting (including the use of low-mounted glare-lights); (b) checkpoint lighting (including lighting within the checkpoint hut); (c) area lighting (these days using relatively low mounting heights and luminaires with accurate beam control); (d) adaptations of standard floodlighting techniques for lighting the faces of buildings and plant; and (e) 'topping-up', (the provision of remotely-placed or adjacent luminaires to 'fill in' areas which would otherwise not be adequately lighted by the remainder of the system). All these techniques, and especially (a) and (b), may be applied empirically, repeating proven standard layouts. Modern practice also includes the use of mobile lighting (14.2) to deal with temporary risks, and is used in hostage situations and sieges involving terrorism, riot control and frontier defence, as well as for the defence of ordinary industrial premises against theft, incendiarism, vandalism etc.

**15.3.6** Help is available for the design and specification of security lighting installations; for example, in the UK, the sales engineers of the electricity boards will usually advise in conjunction with the local Crime Prevention Officer of the police whose guidance will be invaluable. For large, complex or high-risk installations, a security consultant or lighting consultant experienced in this class of work may be employed to advise the user. It will be difficult to compare alternative proposals (which, in the case of security lighting schemes may be very dissimilar) without the initial provision of an Outline Lighting Specification based on specific security objectives agreed with the user and his security advisors[1][45].

**15.3.7** In many exterior installations at factories, the lighting is multifunctional, serving the objectives of parking (15.1), vehicular movement (15.2), as well as amenity and prestige (15.4) in addition to the security function. In many such situations, post-top lanterns (15.2.2) may provide an adaptable method of lighting which can serve all these functions for the lighting of internal roadways and open areas for pedestrian and vehicular movement.

## 15.4 Exterior lighting for amenity and prestige

**15.4.1** The night appearance of industrial premises may be of economic importance, particularly if the prestige of the organization may be enhanced by a good visual impression to be gained from adjacent main roads or passing trains. The exterior lighting might be tailored to produce a pleasant appearance, without in any way detracting from its functional value. The provision of the lighting, especially if there are amenity features such as floodlit trees, gardens etc, can do much to foster public relations with local residents, particularly necessary to counter any strain in those relations due to works noise and traffic movements etc which might be a subject of objection. At factories where there have been complaints about pollution from dark chimney plumes, a spotlight aimed at the head of the chimney will reassure local residents that no dark emissions are occurring in the night hours.

**15.4.2** Where security lighting (15.3) is provided, it may be necessary to adapt the installation so that it does not have a 'prison camp' appearance which might generate local hostility. Roadlighting within the factory perimeter (15.2) might be carefully integrated with adjacent roadlighting e.g. use the same type of lighting equipment, (but not if the public lighting uses *low-pressure* sodium-vapour lamps), and perhaps even providing some gratuitous light on adjacent footpaths as a minor public amenity. Great care must be taken not to direct light in excessive quantity into adjacent private premises, lest this cause nuisance, but it will generally be found that a measure of spill-light out from the factory installation will be acceptable provided it causes no danger by, for example, causing glare to users of nearby roads.

**15.4.3** In these days of energy-consciousness, the application of floodlighting may be condemned as a 'waste of energy', but the security value of such lighting, its amenity value within the factory community, as well as the light which is reflected usefully from the buildings, may provide a basis for justification of this type of lighting—quite apart from the publicity value it may bring. Similarly with amenity lighting for external pedestrian routes between buildings; the lighting for these does not have to be starkly functional, but may be designed to provide a little elegance and colour as well, together with some thought to providing an installation that has an architecturally harmonious appearance by day as well as by night.

Chapter 16

# Calculations for interior general lighting

This chapter and Chapter 17 summarise the essentials of the basic calculations used in lighting design, but are not intended to supplant the book '*Interior Lighting Design*'[7] upon which much that is written here is based. The Lumen Method of lighting design is extensively used, and is doubtless of great utility, though it is by no means the whole technique of how to plan interior lighting. In this chapter the main objective is to familiarise the reader with the concepts and terms used in basic lighting calculations.

## 16.1 The Lumen Method of calculating $E_h$

**16.1.1** The Lumen Method of design is an arithmetical procedure for determining for any desired illuminance in an interior: the number of luminaires of a type; the required lumen output, and hence the power and number of lamps per luminaire. Thus, a lighting layout may be prepared which uses the luminaires correctly (viz at correct mounting height and spacing-to-mounting-height ratio). A further calculation enables the designer to confirm that the resulting Glare Index will not exceed the Limiting Glare Index specified for that application in the *CIBS/IES Code*[5].
**16.1.2** The basic formula is

$$E_h = \frac{l \times n \times N \times UF \times MF \times Abs}{A}$$

where $E_h$ is the illuminance (lux) on a plane, e.g. the floor or a working plane. If not otherwise specified, the working plane is assumed to be 0.85 m above the floor.
$l$ is the number of lumens per lamp. Unless otherwise specified, this will be the Lighting Design Lumens (Appendix III)
$n$ is the number of lamps per luminaire.
$N$ is the number of luminaires.
UF is the Utilization Factor (16.2).
MF is the Maintenance Factor (16.2).
Abs is the Absorption Factor (16.2).

The equation may be transposed to discover any required unknown.
**16.1.3** The minimum number of luminaires which may be used to achieve satisfactory uniformity is determined by the maximum spacing-to-mounting-height ratio ($S{:}H_m$) specified by the luminaire maker. For typical luminaires

(fluorescent tubes and HID lamps) used up to around 6 m, the maximum ratio is usually 1.5:1, viz, if the mounting height above the working plane is 5 m, then the maximum spacing between luminaires will be 7.5 m. For 'high bay' luminaires, typically for use above 5 m mounting height, the ratio is usually 1:1. As one cannot have a fraction of a row, and over-spacing is to be avoided, divide the width (m) by the permitted spacing (m), and round the number up to determine the minimum number of rows. Similarly, divide the length by the permitted spacing and round up the result to determine the minimum number of luminaires per row. Subsequent calculation may reveal that a greater number is required, but there is no objection to somewhat closer spacing; indeed, with fluorescent luminaires, a continuous line of luminaires end-to-end (e.g. on trunking) is economical and good practice. If the manufacturer's constraint on spacing is observed, the resulting uniformity will be not worse than 0.8 viz the ratio of the minimum illuminance to the average illuminance, (though, in some instances it may be necessary to calculate the ratio of the minimum to maximum).

**16.1.4** The spacing of luminaires is governed by how widely the light from the luminaire is spread about the vertical axis. The distribution may be classified as one of the ten standard distributions defined under the British Zonal Classification, these ranging from BZ1 (narrow angle of spread, as used in some high-bay units, and giving a spacing ratio ($S{:}H_m$) of 1:1 or less) up toBZ10 (very wide angle of spread, as used in some dispersive luminaires, giving a spacing ratio ($S{:}H_m$) of 1.5:1 or greater. The BZ classification also affects the Utilization Factor (16.2.1) and the Glare Index (16.3).

## 16.2 Utilization, maintenance, light loss & absorption factors

**16.2.1** The result of performing the Lumen Method calculation (16.1) is to determine the average horizontal illuminance over the area. In practice, not all the lumens emitted by the lamps reach the working plane; some are trapped in the luminaire; some will be emitted towards the walls or ceiling, and only a proportion of these will be reflected back towards the working plane. The proportion of 'useful lumens' actually reaching the working plane is defined by the *utilization factor* (UF):

$$\mathrm{UF} = \frac{\text{Lumens reaching the working plane}}{\text{Total lamp lumens}}$$

The UF may be discovered by reference to tables quoted for actual products by the luminaire makers, or for types of luminaires given in *Interior Lighting Design*[7]. In order to extract the appropriate UF from the tables, it is necessary first to calculate the *Room Index* (RI). The RI takes account of the physical proportions of the room, thus:

$$\mathrm{RI} = \frac{LW}{H_m(L+W)}$$

where $L$ is length of room (m)
$W$ is width of room (m)
$H_m$ is mounting height of the luminaires above the working plane.

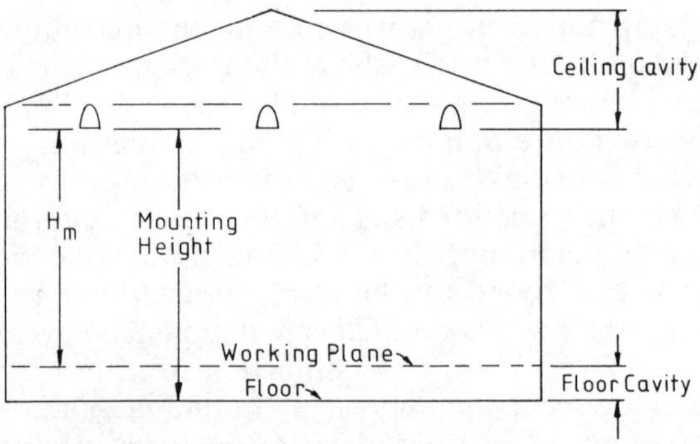


*Figure 16.1* Vertical dimensional parameters in lighting calculations.

The other information needed to enter a table of UFs is the reflection factor of the walls, ceiling cavity and floor cavity (Figure 16.1).

**16.2.2** The way the light flux is distributed from the luminaire affects the maximum permissible spacing if the requirements of uniformity of illuminance are to be met (16.1.3). The distribution also affects the Utilization Factor; generally, a narrow angle of distribution will project the light strongly downward, so that little light goes directly to the walls, resulting in a high UF. Conversely, wider distributions (e.g. those with bigger BZ numbers) will tend to direct more flux towards the walls, and thus the UF will tend to be smaller. Note that the UF is determined by the interreaction of the room proportions, e.g. the Room Index (16.2.1) and the luminaire distribution, e.g. the BZ number (16.1.4). It will also be shown that the BZ classification affects the glare effect of the installation (16.3). Light going to the working plane directly from the luminaire is termed the *direct component,* while that which reaches the working plane after one or more reflections from building surfaces is termed the *indirect component.* In installations having a high proportion of indirect component, if the walls etc are strongly coloured, the light arriving at the working plane may be degraded as regards colour to a greater or lesser extent—a matter of some importance in rooms where fine colour work is to be performed, or where the rendering of the colours of objects in the space is required to be accurate (4.3). The value of having light-coloured walls and ceilings in a factory is considerable, no matter what type of light distribution is given by the luminaires; if the BZ number is high, light coloured walls will reflect light—some of which will find its way usefully to the working plane; if the BZ number is low, then it becomes important that the walls are of light colour to reflect what little light reaches them and to lessen the otherwise sombre effect that may be created. The combination of dark walls and low BZ number luminaires can be quite depressing for even if the illuminance on the working plane is of the order of thousands of lux, the overall appearance of the room will be gloomy. Similarly, if the ceiling cavity (Figure 16.1) is of dark colour, the appearance of the room may have what is termed *tunnel effect* where the occupants feel that they are in a low-ceilinged tunnel, with the luminaires set into its roof—even if, in fact, there is plenty of headroom above the luminaires.

**16.2.3** The Maintenance Factor (MF) used in the Lumen Method calculation (16.1.1) provides an allowance for the light lost due to dirt on the lamp and luminaire. It is defined as:

$$MF = \frac{\text{Lumens emitted by the luminaires when soiled}}{\text{Lumens emitted by the luminaires when clean}}$$

The light output per luminaire will decline from the time of installation due to the gradual build-up of dirt, and will be substantially restored by thorough cleaning. The actual MF will depend upon the rate of soiling which must be estimated from tables[40], or the estimate may be based on the soiling record of similar luminaires in the same location or in a location having the same propensity for soiling, and the frequency of cleaning. In typical industrial locations the frequency of cleaning is likely to be not greater than once every six months except in exceptionally dirty locations which may justify quarterly cleaning. Typically, annual cleaning will be carried out, and it is only the exceptionally clean locations which will justify cleaning the luminaires at 2-yearly intervals. Because all possible use should be made of available daylight (both for energy-saving and amenity reasons) (2.3), and because dirty walls and other room surfaces will also absorb light (and thereby waste energy), cleaning programmes should include cleaning of all glazing in walls and ceilings at about the same frequency as the cleaning of the luminaires. Wall and ceiling surfaces in industrial locations will need to be cleaned at least annually, with repainting at suitable intervals.

**16.2.4** The economic effect of using light-coloured room surfaces (16.2.2) and attending to regular cleaning of the luminaires and the room surfaces (16.2.3) can be significant. Taking as an example a medium-sized industrial area having a Room Index (16.2.1) of 2, if when using an installation comprising fluorescent tubes in plastic open base troughs, the walls and ceiling had reflection factors of 0.5 and 0.7 respectively, the UF would be around 0.62. But if the reflection factors of the walls and ceiling should be of 0.1 and 0.3 respectively, then the UF would be only 0.5. In other words, for the same energy consumption, the illuminance would fall by 23 per cent. Conversely, raising the reflection factors of the interior would enable a given illuminance to be achieved with a consumption of only three-quarters of the energy.

**16.2.5** The economic value of a planned maintenance routine is often under-estimated (13.1). The annual average rate of depreciation of lumen output from luminaires (*D*) is found from the formula

$$D = \frac{E_0 - E_1}{E_0 \times T} \qquad (1)$$

where $E_0$ is the initial illuminance
$E_1$ is the illuminance after a period $T$, and
$T$ is the time in years.

Luminaires do not all depreciate at the same rate; ventilated luminaires which have a through-draught may have a *D*-factor as low as 0.05, while a non-ventilated reflector or an enclosed luminaire which aspirates soiled air might have a *D*-value around 1.00. The *optimum economic cleaning interval* (*T*) is that period over which the cost of the light lost by dirt equals the cost of cleaning the luminaires, and is given by

$$T = \frac{-C}{A} + \sqrt{\frac{2C}{A \times D}} \text{ years} \tag{2}$$

where $A$ is the annual cost of operating the luminaires (i.e. energy consumption, amortisation, lamp replacements etc) without the cleaning cost,
$C$ is the cost of cleaning the luminaires once,
$D$ is the average rate of depreciation of lumen output due to soiling (Equation (1)).

**16.2.6** In calculating the economic cleaning interval $T$ (Equation (2), (16.2.5), it must be appreciated that in any period the rate of soiling of a luminaire is not uniform. The rate of soiling seems to be fastest in the initial months, and thereafter the soiling is at much slower but fairly uniform rate (Figure 16.2). Referring to the figure, one would expect most open ventilated industrial luminaires to have D-values of Classes B to E in normal industrial situations, and E to H in exceptionally polluted atmospheres. The curves indicate that shortening the cleaning interval from one year to six months has only marginal effect in the former cases, but the lumen saving is greater for the latter cases.

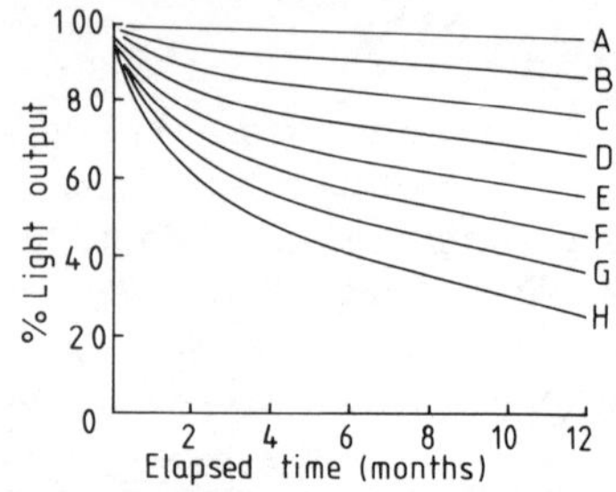


*Figure 16.2* Luminaire categories of depreciation[40].

The accumulative effect of periodic cleaning and re-lamping at appropriate intervals is discussed in paragraph 13.3.5, and is illustrated in Figure 13.2. The times of cleaning of luminaires and periodic re-lamping should be adjusted so that cleaning and re-lamping takes place as one operation with a saving in labour cost. The designer is faced with certain choices; for example, he may decide to clean more frequently and thus need to install a smaller number of lumens to achieve a required average-through-life illuminance; or he could decide to clean less frequently (because, for example, access to the luminaires is difficult, costly or dangerous), and therefore have to install a greater quantity of lumens to achieve the required average-through-life illuminance. The latter course will be more wasteful of energy, but could be justified on economic grounds in some cases.

**16.2.7** In calculating by the Lumen Method, no allowance need normally be made for the decline in lumen output due to the lamps ageing. The calculation is normally for the *service illuminance*, i.e. the mean illuminance throughout the maintenance cycle of an installation and averaged over the relevant area. In the calculation, it is customary to use the *Lighting Design Lumens* figure issued by the lamp-maker, this (for discharge-lamps) usually being based on the lumen output at 2000 h. If required, this figure can be substituted with the *initial lumens* (i.e. the lumen output obtained from a new lamp), or with the *terminal lumens* (i.e. the lumen output to be obtained from a lamp at the end of its nominal life (but, see paragraph 13.3.5) to calculate the lowest average illuminance that will be produced by the installation.

**16.2.8** The Maintenance Factor normally employed is as defined in paragraph 16.2.3, viz it is ratio of the lumen outputs of the luminaires in the clean and soiled conditions, and normally assumes that the calculation will be used with the lamp lighting design lumens (16.2.4). This is anomalous, and obviously has a small inherent inaccuracy, for, by the time the luminaires have become soiled, the lamps will have aged to some degree. In the CIE system[47] this is overcome by introducing the concept of the *Light Loss Factor*. This is a factor which takes into account lamp depreciation as well as the light loss due to soiling of the luminaires and room surfaces. In the practical application of the CIE Light Loss Factor there may be difficulty in obtaining reliable figures for initial and end-of-life lumen outputs of lamps, but, none-the-less, given the nominal Lighting Design Lumens figure for the lamps, a very good approximation of actual lighting conditions may be obtained (Table 9). Table 9 may be employed (or the full calculation performed as indicated in the *CIBS/IES Code*[5]) to give the following: (a) to measure the illuminance from a new lighting installation (at 100 h) to confirm the general accuracy of the calculation, and (b) to measure the illuminance over a period of time to confirm the accuracy of the Maintenance Factor used in the calculation.

**Table 9** Typical light loss factors, based on table in 1977 Edition of CIBS/IES Code[5]

| *Recommended illuminance based on:* | *Lamp lumens* | *Lamp light loss factor* | *Room category* | *Luminaire and room surface light loss factor* | *Total light loss factor (LLF)* |
|---|---|---|---|---|---|
| | | | clean | 0.85 | 0.7 |
| Minimum illuminance | Initial lumens (100 h) | 0.8 | average | 0.75 | 0.6 |
| | | | dirty | 0.6 | 0.5 |
| | | | clean | 0.85 | 0.85 |
| Minimum illuminance | End-of-life lumens (70% rated life) | 1.0 | average | 0.75 | 0.75 |
| | | | dirty | 0.6 | 0.6 |
| | | | clean | 0.9 | 0.8 |
| Service illuminance* | Initial lumens (100 h) | 0.9 | average | 0.8 | 0.7 |
| | | | dirty | 0.7 | 0.6 |
| | | | clean | 0.9 | 0.9 |
| Service illuminance* | Lighting design lumens (2000 h) | 1.0 | average | 0.8 | 0.8 |
| | | | dirty | 0.7 | 0.7 |

* Service illuminance: see Paragraph 16.2.7 and Appendix II.

**16.2.9** *Absorption Factor* is a factor used in Lumen Method calculations (16.1.1) to allow for the light loss due to absorption and scattering of light during its passage from the luminaires to the working plane. This is a factor that is often completely omitted from calculations, or, if employed, is often much under-estimated. From the author's work in studying the absorption effect of atmospheric moisture in exterior lighting installations[1] it has become apparent that even a small degree of atmospheric pollution in a high-roofed building can not only cause light loss due to absorption, but that the scattering or diffusing effect may be such as to change the nett effect of the luminaire light distribution to the equivalent of raising its BZ Number (16.2.2) by one or two whole units, with the result that the effective Utilization Factor is very much reduced. In modern lighting practice, this effect is often completely overlooked, so that in high-roofed factory buildings there may be serious under-lighting in the presence of relatively small degrees of pollution by steam, dusts, atmospheric moisture etc. As there seems to be no accurate way of forecasting the proper Absorption Factor (Abs) to be used, it can only be recommended that experiments are conducted in the actual location (or a closely-similar one in the case of new structures in which the plant is not yet operating), to measure the illuminance under clear-air and polluted conditions and to calculate the ratio between the two readings. To give an appreciation of the magnitude of the light loss that can occur, measurements made in a metal stamping shop on two successive nights showed a 13 per cent difference in illuminance, the change in reading being accounted for by an almost imperceptible degree of mist that percolated into the building on one of the nights from an adjacent canal. The illuminance measured in another location (a drop forge) showed a reduction in illuminance of 24 per cent between clean conditions (before work started) and polluted conditions (after six hours production). The absorption loss in a food factory kettle room was estimated at around 12 per cent, accurate measurement not being possible because of the diffusing effect due to steam condensing on the enclosed luminaires. In a very badly polluted foundry, the absorption loss was estimated at around 25 per cent during pouring, and in this case the soiling of the luminaires was visible within one shift of cleaning, and probably accounted for a substantial part of the measured loss.

## 16.3 Calculation of direct glare

**16.3.1** The subjective sensation of *glare* (1.1.7) is capable of being evaluated and represented by a non-quantitative number that expresses the degree of discomfort. For interior lighting installations in which the luminaires are arranged in regular patterns, it is possible to calculate a *glare index* for any real or proposed design layout, and to compare this with the *limiting glare index*, e.g. the recommended maximum glare index for that location[5]. Such glare indices take account only of the *direct glare* produced by the installation, i.e. that due to the luminance of the luminaires. So far, no practical instrument has been developed which enables the measurement of glare in actual installations, but there is a fair degree of agreement between observers who can make a subjective assessment of glare conditions in an interior. The basis of the British Glare Index System was devised by the Building Research Station out of data produced by some hundreds of observers who visited a large number of installations and were required to record their assessments of the glare conditions as being either 'just imperceptible', 'just perceptible', 'just tolerable' or 'just intolerable'. From this data, numerical representations of the group opinions were devised.

**16.3.2** The Glare Index is a number lying between 10 and 30 which represents the degree of glare sensation, 10 being 'virtually no preceptible glare', and 30 being 'unbearably glaring', and with glare indices in the range 16 to 22 being those commonly recommended as the Limiting Glare Indices in industrial locations. Where fine work is being performed, reduction in glare will materially improve performance of the operators and reduce their fatigue, especially if the work is of long duration. But in the performance of coarse tasks, a glare index in the range 22 to 28 may well be justified if reduction would be costly, for any task performance improvement is likely to be marginal and the fatigue reduction notional, especially if the exposure to the glare is of short duration. It is necessary to keep a sense of proportion in this matter; striving for very low glare indices is not usually of any marked economic value and very glare-free environments tend to be literally lack-lustre and very uniform to the point of being visually boring. On the other hand, we can tolerate very high glare indices for short periods, e.g. driving at night on unlighted roads and having to cope with the headlights of oncoming vehicles. However, where glare will materially affect performance (e.g. output rate, quality etc) or safety (e.g. by masking dangers, preventing adequate vision of operators or drivers) then steps must be taken to design lighting having reduced direct glare. In many industrial installations, the important glare will be that reflected from specular surfaces, wet surfaces etc, so that luminaires will have to be positioned to reduce the incidence of these reflections. Very often the best way to reduce the glare sensation from reflections will be to use large-area low-brightness luminaires or overall luminous ceilings; but, these days, the use of polarized light enables the brightness of reflections in selected angles of view to be minimized (Appendix V), a technique used in inspection and other fine work (3.4.1 (h)).

**16.3.3** In interior lighting installations, the glare sensation is related to the brightness of the glare source, the 'apparent size' of source (combination of actual size and distance), the brightness of the background against which the glare source is viewed, and a positional factor relating the position of the glare source to the direction of view. These factors are combined to calculate the Glare Index[(7)]. The BRS Glare Index does not take into account the effect of time; but, in many real situations, the time to adapt to a change in field brightness is important (4.3.7). This is experienced on moving between areas lighted to different illuminances, and particularly when passing between outdoor and indoor areas, both by night and by day (2.3.2). But, the Glare Index system is concerned only with the steady condition, where the subject is within the lighted area, and—presumably—has adapted to the general field luminance. An important factor in reducing glare in interiors is the brightness of the background against which the luminaires are seen; thus, the provision of light colours in the ceiling cavity (4.1, 16.2.1) will materially reduce the glare experienced from any type of luminaire, and the more so for luminaires which allow a proportion of upward light. Where the floor, or bench-tops etc, are of light colour, light will be reflected to the ceiling cavity, which, if of good reflectance, will become of sufficient luminance to reduce the glare effect from the luminaires. This point is made to emphasise that luminaires by themselves cannot be 'glare free' or of 'low glare type'; it is the combination of luminaire and interior that determines the degree of direct glare experienced. Thus, although luminaires exercising greater control over the emergent flux, viz those with low BZ Numbers (16.1.4), will often be those selected for designs intended to produce low glare indices, it is certainly possible to produce very uncomfortable results with such luminaires if the general pattern of luminances within the interior are badly engineered. The methods of calculating glare indices take this into account (16.3.4).

**16.3.4** No useful purpose would be served in taking the reader through the mathematical steps to compute a glare index, for this is adequately explained in the book 'Interior Lighting Design'(7) to which the reader is recommended. It will suffice to state that a very good approximate method is given therein, which, even in the absence of complete photometric data from the luminaire manufacturer will enable the possibility of unacceptable glare conditions to be predicted by simple calculations based on the dimensions and lumen output of the luminaire type, the room dimensions and its decor. The Limiting Glare Indexes recommended in the *CIBS/IES Code*(5) (which are reproduced also in 'Interior Lighting Design'(7)) were empirically decided on the basis of the extensive experience of the compilers.

**16.3.5** It is common experience that light reflected from glossy surfaces tends to mask the objects thereon—e.g. the difficulty of reading print on glossy paper when a bright image of a luminaire is reflected from the paper surface. This difficulty can be reduced by reducing the brightness of the potential glare source in the necessary directions (16.3.2). A method of describing the visibility under such conditions is the use of the *Contrast Rendering Factor* (CRF). It should be noted that some luminaire distributions which have been engineered to reduce direct glare and enhance the illumination of vertical surfaces (e.g. 'batswing' and 'trouser-leg' distributions (17.3)) may produce a worse CRF. In locations where highly reflecting surfaces abound, for example, in sheetmetal works, it will be almost impossible to control both direct and indirect glare in all directions of view, and the designer will have to make compromises. However, the use of polarized light (Appendix V) may enable very good results to be obtained, though with some extra cost compared with conventional lighting systems.

## 16.4 Approximations and calculation aids

**16.4.1** No approximation can give more than a general guide in lighting design, for the number of variable factors is great in every scheme. Further, guide tables limit the choice of lamps, luminaires and illuminances that may be employed. None-the-less, they have their uses, particularly in the early stages of considering a proposal to carry out new lighting work. Table 10 gives a guide to the number of fluorescent luminaires to provide 500 lux in typical rooms, while Table 11 gives a guide to the number of 250 W or 400 W SON lamps to provide 300 lux or 500 lux in typical bays. It is stressed that these tables should not be substituted for full calculations. A wider range of illuminances is provided for in the *IES/CIBS Code* (See Appendix II).

**Table 10** (See 16.4.1) A guide to determining the approximate number of fluorescent lamp luminaires needed in rooms of common sizes to achieve approx. 500 lux.

The luminaires will be twin-tube, housing either two 38 mm 65 W or two 26 mm 58 W Colour 84 or WHITE fluorescent tubes of Lighting Design Lumens around 4800 lumens per tube. Mounting height of 2.3 to 3.0 m above floor, with the tubes evenly spaced over the area. (Maximum spacing to be one-and-a-half times the nett mounting height between the luminaires and the working plane—here assumed to be 0.85 m above floor.) Calculations assume average reflectances and cleanliness of decor, and use of enclosed prismatic controllers; luminaires ceiling-mounted.

| *Room width (m)* | *Room length (m)* | | | | | | | |
|---|---|---|---|---|---|---|---|---|
| | *3* | *4* | *5* | *6* | *7* | *8* | *9* | |
| 3 | 2 | 2 | 2/3 | 3 | 3/4 | 4 | 4 | For larger |
| 4 | | 2/3 | 3 | 4 | 4 | 4/5 | 5 | rooms, allow |
| 5 | | | 4 | 4/5 | 5 | 6 | 6/7 | 1 luminaire |
| 6 | | | | 5 | 6 | 7 | 7/8 | per 8 $m^2$, |
| 6 | | | | | 7 | 8 | 8/9 | evenly spaced |

*Method of use*

The luminaires are to be arranged in rows parallel to the room axis.

To determine the number of rows, divide the room width by 1.5 × the mounting height above the working plane, and round up the result. Divide the room width by the number of rows to determine the spacing between the rows. The distance from the outer rows to the walls should be half the spacing between the rows.

Divide the recommended number of luminaires by the number of rows and round up, thus determining the number of luminaires per row. Divide the room length by the number of luminaires per row to determine the spacing centre-to-centre of luminaires in the rows. The distance from the centres of the end luminaires to the end walls should be half this distance.

* This table is based upon data issued by Philips Lighting for their 'Powerslimmer Commercial Kombipak' 2 × 58 W fluorescent tube luminaire.

**Table 11*** (See 16.4.1) A guide to determining the approximate number of either 250 W or 400 W Type SON lamps in high-bay luminaires to achieve either 300 lux or 500 lux.

In each of the following tabulations, the upper number is the number of rows of luminaires, and the lower number is the spacing along the rows in metres.

*(a) Using 250 W lamps to provide 300 lux*

| *Bay width (m)* | *Mounting height (m)* | | | | | |
|---|---|---|---|---|---|---|
| | *6* | *7* | *8* | *10* | *12* | *15* |
| 10 | 2<br>7.8 | 2<br>8.0 | 2<br>7.2 | 2<br>8.0 | 2<br>7.2 | 2<br>6.5 |
| 12.5 | 2<br>7.2 | 2<br>7.2 | 2<br>6.5 | 2<br>6.5 | 2<br>6.0 | 2<br>5.5 |
| 15 | 2<br>6.0 | 2<br>6.0 | 2<br>6.0 | 2<br>5.5 | | |
| 17.5 | 3<br>7.8 | 3<br>8.0 | 3<br>8.0 | | | |
| 20 | 3<br>7.2 | 3<br>7.2 | 3<br>7.2 | | | |
| 22.5 | 3<br>6.5 | 3<br>6.5 | 3<br>6.0 | | | |
| 25 | 4<br>7.8 | 4<br>7.2 | 4<br>7.2 | | | |

*(b) Using 400 W lamps to provide 300 lux*

| *Bay width (m)* | *Mounting height (m)* | | |
|---|---|---|---|
| | *10* | *12* | *15* |
| 15 | | 2<br>9.0 | 2<br>9.0 |
| 17.5 | 2<br>9.0 | 2<br>8.0 | 2<br>8.0 |
| 20 | 2<br>8.0 | 2<br>7.2 | 2<br>7.2 |
| 22.5 | 3<br>10.3 | 3<br>10.3 | 3<br>10.3 |
| 25 | 3<br>9.0 | 3<br>9.0 | 3<br>9.0 |

*(c) Using 250 W lamps to provide 500 lux*

| *Bay width (m)* | *Mounting height (m)* | | | | |
|---|---|---|---|---|---|
| | *6* | *7* | *8* | *10* | *12* |
| 10 | 2<br>5.1 | 2<br>5.1 | 2<br>4.8 | 2<br>4.5 | 3<br>6.0 |

*(d) Using 400 W lamps to provide 500 lux*

| Bay width (m) | Mounting height (m) | | | | | |
|---|---|---|---|---|---|---|
| | *6* | *7* | *8* | *10* | *12* | *15* |
| 10 | | | | | | 2<br>7.2 |
| 12.5 | 2<br>8.0 | 2<br>8.0 | 2<br>8.0 | 2<br>7.2 | 2<br>6.5 | 2<br>6.0 |
| 15 | 2<br>7.2 | 2<br>7.2 | 2<br>6.5 | 2<br>6.5 | 2<br>5.5 | 2<br>5.5 |
| 17.5 | 2<br>6.0 | 2<br>6.0 | 2<br>6.0 | 3<br>8.0 | 3<br>7.2 | 3<br>7.2 |
| 20 | 3<br>8.0 | 3<br>8.0 | 3<br>8.0 | 3<br>7.2 | 3<br>7.2 | 3<br>6.2 |
| 22.5 | 3<br>7.2 | 3<br>7.2 | 3<br>7.2 | 3<br>6.5 | 3<br>6.5 | 3<br>6.0 |
| 25 | 3<br>6.5 | 3<br>6.5 | 3<br>6.5 | 3<br>6.0 | 3<br>6.0 | 4<br>7.2 |
| 30 | 4<br>7.2 | 4<br>7.2 | 4<br>7.2 | 4<br>6.5 | 4<br>6.5 | 4<br>6.5 |

*Method of use*

Divide the bay width by the upper number to determine the spacing between the rows. The distance from the outer rows to the walls should be half this distance.

Divide the bay length by the lower number, and round up to determine the number of luminaires per row. Divide the bay length by the determined number of luminaires per row to calculate the spacing between the luminaires in the rows. The distance from the end luminaires to the end walls should be half this distance.

* This table is based upon data issued by Philips Lighting for their 'Powerslimmer High Bay Kombipak' 250 W and 400 W high bay SON luminaires.

**16.4.2** The procedure for performing a Lumen Method design calculation is given fully in the book '*Interior Lighting Design*' which takes the reader through the simple arithmetical procedure step by step. It also deals with the calculation of the glare index which will relate to each proposed lighting layout. The Lumen Method is summarised in this Chapter (16.1). Where many calculations are to be performed, a computer may be used. Programmes exist to evaluate not only the relative merits of many alternative layouts, but also to give costs and running costs for the alternatives during the design process, a useful means of eliminating those schemes which are technically feasible but which would be uneconomic in use. At a much more elementary level of mechanization, calculations by the Lumen Method and for Glare Index may be performed on a pair of special circular slide-rules known as 'Mear's Calculators'.†

†Messrs M. H. Mear & Co., 56 Nettleton Road, Dalton, Huddersfield, Yorks.

Chapter 17

# Directional lighting

As in the preamble to Chapter 16, it is pointed out that this chapter is intended only to familiarise the reader with the main points about the calculation of directional lighting, and it is to the book *'Interior Lighting Design'*[7] that the reader should refer for more complete information. Scientific definitions of the terms are given in the *CIBS/IES Code*[5].

## 17.1 Point-by-Point Method of calculating E

**17.1.1** This type of calculation is based upon the *Inverse Square Law*, and applies only to a source which is a 'virtual point source', (i.e. having negligible area in relation to the distance over which the calculation is performed) ignoring the effects due to reflection of light (e.g. for building surfaces or nearby objects). The relationship is

$$E = \frac{I}{d^2}$$

where $E$ is the illuminance (lux),
$I$ is the source intensity in the direction under consideration (candelas), and
$d$ is the distance between the source and the point of measurement (metres).

The illuminance $E$ decreases as the square of the distance, and the illuminance is distributed over an area that increases as the square of the distance (Figure 17.1).

**17.1.2** The brightness of a source does not diminish with increase of

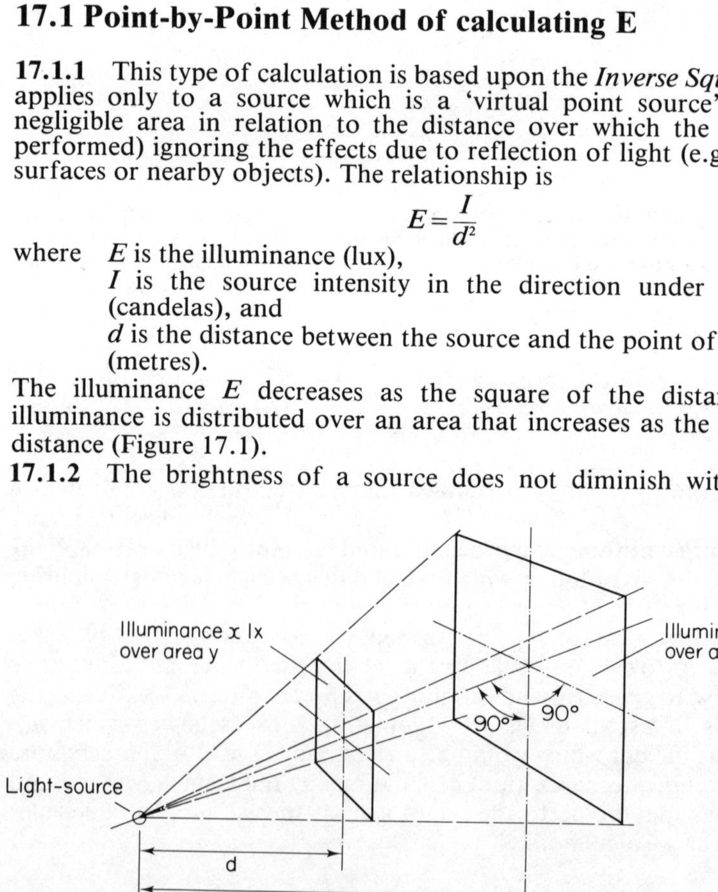


*Figure 17.1* Inverse Square Law of illumination.

distance, but is intrinsic to the source. However, the source as perceived becomes smaller with distance according to the Inverse Square Law, and may therefore appear to be less glareful.

**17.1.3** The relationship between illuminance, intensity and distance given in Paragraph 17.1.1 holds good only if the plane of measurement is normal to the incident beam. When the beam strikes the plane of measurement at any other angle $\theta$, the correction cos $\theta$ must be applied

$$E = \frac{I}{d^2} \cos \theta \tag{1}$$

Because it is difficult in practice to measure $d$, and the horizontal distance $D$ and the height of the source $H$ are more readily ascertained, the formula may be rearranged to calculate the horizontal illuminance $E_h$ thus

$$E_h = \frac{I \cos^3 \theta}{H^2} \text{ or } E_h = \frac{I}{D^2 + H^2} \cos \theta \tag{2}$$

The illuminance normal to the beam may be calculated by

$$E_n = \frac{I \cos^2 \theta}{H^2} \tag{3}$$

and the vertical illuminance is given by

$$E_v = \frac{I \cos^2 \theta \sin \theta}{H^2} \tag{4}$$

These relationships are shown in Figure 17.2

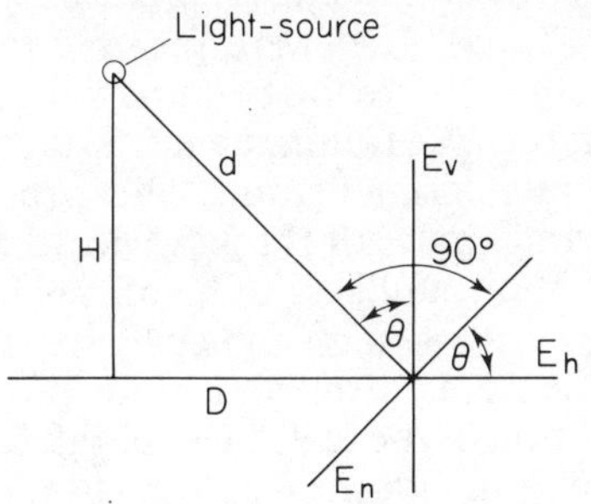


*Figure 17.2* Cosine Law of illumination.

## 17.2 Vertical, cylindrical and spherical illuminance

**17.2.1** Vertical illuminance may be measured with the light-cell of the lightmeter vertical. Note that the measurement is in a vertical plane of a single orientation. Although conventional lighting uses the Lumen Method (16.1) to determine the illuminance on the horizontal working plane, in many practical situations the working plane is not horizontal, but may be vertical or angled, e.g. drawing boards, blackboards, or work on the vertical side of a machine etc.

**17.2.2** It has been stated that a measurement of vertical illuminance with a conventional lightmeter is in only one plane of orientation (17.2.1). Here one can introduce the concept of imagining a small cylinder with its axis vertical placed at the point of measurement; if it was possible to measure the illuminance on all parts of that cylinder, we would have computed all the vertical illuminance contributions from all directions in plan, and this would be termed the *mean cylindrical illuminance.* Thus, when discussing the flow of light on to vertical surfaces in any plane we could express the relative magnitude of the cylindrical illuminance ($E_c$) to the horizontal (planar) illuminance ($E_h$) as the cylindrical/planar ratio ($E_c/E_h$) and this would provide a useful measure of how well vertical surfaces would be illuminated generally in comparison with the horizontal illuminance. In practice, four measurements of vertical illuminance made at a point, with the light-cell rotated through 90° between measurements, and the four readings averaged, would give a very good approximation to the mean cylindrical illuminance. However, special photocells which compute the arriving light from all directions in the vertical planes are available to do the job more accurately.

**17.2.3** If we now introduce a further concept, the reasoning that lead to the explanation of the mean cylindrical illuminance (17.2.3) can be extended. Imagine now that a small sphere is located at the point of measurement, and that it is possible to summate and average all the lumens arriving on the surface of that sphere; this would give the mean spherical illuminance or *scalar illuminance* at that point. This would give a useful measure of the flow of light to that point, but would tell you nothing about which parts of the sphere was the best lit. However, a very good approximation of the scalar illuminance at a point can be made by six photocell measurements, turning the cell so that is directed to the six directions that would be represented by the faces of a small cube at that point, i.e. four vertical measurements rotating the cell through 90° between measurements, one with the cell facing upward, and one with cell facing downward; these six readings averaged will give the approximate scalar illuminance at that point.

**17.2.4** A further study of the way light arrives at a point is possible by extending the concept of the small sphere (17.2.4). If it were possible to measure to discover which parts of the small sphere were the best and worst lit, then the line between them would be the *vector* of illuminance at that point. This is a matter of some importance in exterior lighting where the engineer is concerned with mounting heights and angles of throw to project light great distances[(1)] but, in the author's opinion, unlikely to be of much practical application for most interior lighting systems. It is claimed by some workers that an accurate appreciation of the revealment and modelling effects of lighting can be obtained by consideration of the scalar-vector ratio, but how this can be applied to practical problems in industrial lighting is not clear; perhaps the ratio of mean cylindrical illuminance to planar illuminance ($E_c/E_h$) would be more helpful for ensuring that light will come in sufficient quantity to illuminate all sides of a large machine in the middle of a working area, for example.

## 17.3 Designing for enhanced vertical illuminance

**17.3.1** Valuable as they are to the proper study of illuminating engineering, the special measures of illuminance (vertical, cylindrical, scalar and vector), are not used very much in practical design work unless one has a suitably programmed computer at hand, and is in possession of a great deal of data about the luminaires which may be used. To bring the concepts into a practical framework, let us state simply that measurement of the horizontal illuminance is still a good measure of the quantity of light in an interior; but, any steps which can be taken to improve the mean cylindrical illuminance will probably improve visual comfort and certainly will enhance the illuminance of vertical surfaces in the interior. To achieve enhancement of the $E_c/E_h$ ratio in an interior, the following steps could be taken:

(a) Use the type of luminaire that has the biggest possible BZ number (e.g. widest angle of distribution) compatible with suitable control of glare. (This will not exclude the use of batwing or trouser-leg distributions if such luminaires are appropriate and available.)

(b) Employ a type of luminaire that gives some upward light; this will not only help overcome tunnel effect but will help keep the luminaire clean if it is a through-vented pattern (11.1.7).

(c) Use the lightest possible colours for ceiling, walls and floor, and institute a programme of regular cleaning and periodic redecorating of those surfaces. Also, use light colours for furniture, machines etc.

The result of these steps will be to cause a greater integration of light within the interior, so that by re-reflection light will come from many directions to each point in the working area, viz there will be an improvement in the $E_c/E_h$ ratio. A word of caution; by the application of light colours, do not create a featureless *bland field* (4.1.1).

**17.3.2** In addition to enhancement of vertical illuminance by the measures suggested (7.3.1), there is always the possibility of using local lighting to provide preferential illumination in any desired direction according to the features of the visual task (2.2). Control of vertical illuminance can also be effected by positioning and orientation of luminaires. For example, the attenuation of light from overhead luminaires reaching into shelf cavities, and to the lowest shelves in a store or library is marked because of the cosine effect (17.1.3); a practical point to take note of when lighting shelving facing into a narrow aisle, is that better illuminance at the lowest shelves (and better penetration into shelf cavities) will be obtained by placing fluorescent tubes transversely rather than parallel to the aisle faces (Figures 17.3). A further practical point is that it will significantly improve lighting conditions under these circumstances if all the shelving is painted white (including undersides of shelves), and if the floor is painted with white floor-paint, e.g. epoxy-resin paint. Reflections of light will improve the vertical illuminance on the face of the stacks.

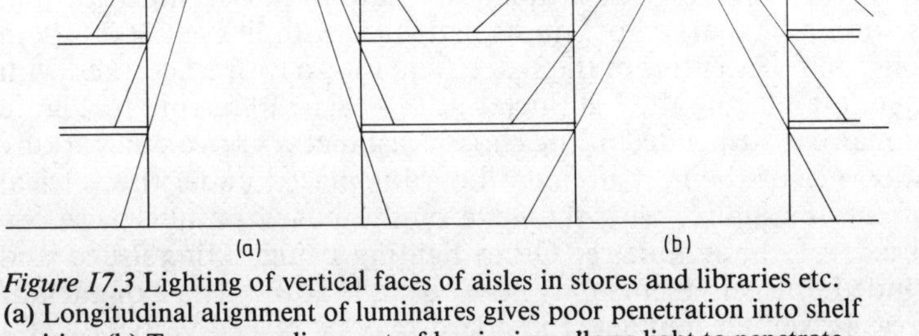


*Figure 17.3* Lighting of vertical faces of aisles in stores and libraries etc. (a) Longitudinal alignment of luminaires gives poor penetration into shelf cavities. (b) Transverse alignment of luminaires allows light to penetrate better into shelf cavities.

**17.3.3** In larger areas, the steps which may be taken to improve the vertical illuminance include the use of 'batwing' or 'trouser-leg' distribution luminaires (Figure 8.1) which are designed to restrict the illuminance due to light at the near-vertical angles and distribute the lumens preferentially at an angle roughly between 40° and 70° from the downward vertical axis. This distribution, while giving restricted brightness above the 70° line (which would tend to cause direct glare) preferentially lights vertical surfaces which are presented to the directions of adjacent luminaires. Another method of improving the illuminance of vertical surfaces in an area is to use cross-lighting (Figure 8.1) by wall-mounted luminaires, or, in larger areas, perhaps mounted at the heads of stanchions. Where the desired improvement in vertical illuminance needs to be mainly in one plane, then angled or parabolic reflectors may be used, positioning them carefully to avoid causing excessive discomfort glare to persons who must work facing them.

## 17.4 Practical method of producing design aids

**17.4.1** In practical lighting design work, sophisticated calculations are but rarely employed, for it is usually more convenient to read off values from nomograms[(7)] or to employ *isolux diagrams*. For task lighting studies, the author has found it worth while taking the trouble to measure the illuminance from fluorescent luminaires at various mounting heights and distances from vertical planes, and to chart these in the form shown in Figure 17.4. Similarly, for spotlights and other luminaires giving strong directional beams, it is a simple matter to chart out the actual measurements taken at a few likely distances and angles as shown in Figure 17.5. Failing such practical steps being possible, if photometric data on the selected luminaire is available, then Point-by-Point calculations may be performed. It should be mentioned that at one time lighting manufacturers customarily published the full data about their luminaires, but the practice is falling into disuse; however, full photometric information is usually available on request, usually in the form of a computer print-out, for it is customarily by the computer that these calculations are performed. For those in industry who have the task immediately before them without full photometric data, the time and trouble to set up and measure the characteristics of a few chosen luminaires under a selected range of distances as shown in Figures 17.4 and 17.5 will be well repaid by the production of sound practical data that can immediately be applied to problems on the shop floor.

**17.4.2** A simple method of producing useful isolux diagrams from a fluorescent luminaire to be used for local lighting is by the use of a measurement board consisting of ordinary hardboard measuring 1200 mm square. This should be marked off on its matt face with lines at 300 mm in both directions, and the centres of the sixteen squares so formed marked with a cross to facilitate placing the lightmeter cell in each. Referring to Figure 17.4, the luminaire is suspended at the chosen distance $h$ above the working plane, and a corner of the measurement board is placed under the vertical centreline of the luminaire, with the edge of the board parallel with the longitudinal axis of the luminaire. Other lighting being extinguished and daylight excluded, measurements may now be made at the centre of each of the 16 squares. In similar manner, the board may be placed vertically at a chosen distance $d$ off the centreline of the suspended luminaire, and a similar

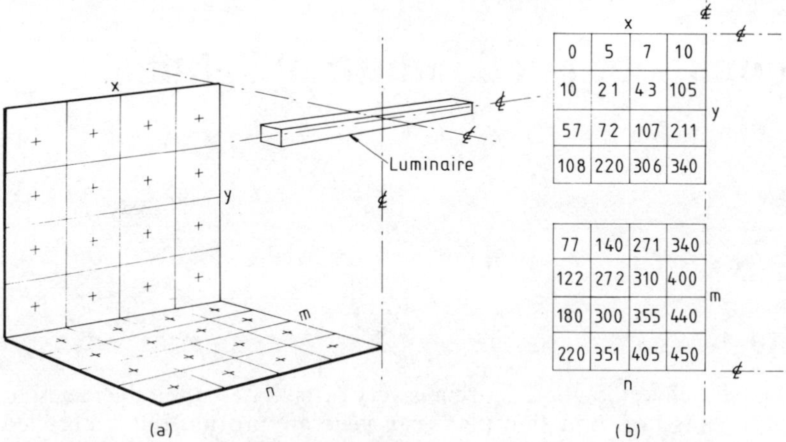


*Figure 17.4* Use of measurement-boards to record illuminances in studies of local lighting. (a) Board used in vertical and horizontal planes; measurements taken at centres of squares. (b) Typical results (example only, not for use).

series of measurements taken. As these luminaires are symmetrical in their output, in effect only one quarter of the possible measurements are taken. The procedure may be repeated for several different values of $h$ and $d$ as required, and the resultant charts retained for future use.

**17.4.3** A practical method of producing an isolux diagram from spotlights and other strongly directional lamps used for task lighting is again to use a measurement board as described (17.1.5). For small spots of light, aim the beam at the centre of the board; for larger ones aim at one corner of the board (for the other four quadrants will be identical if the beam axis is normal to the board) (Figure 17.5). For angles $\theta$ other than 90° a similar procedure is followed, but noting that if the aim is at a corner as explained above, then the resulting diagram will be symmetrical only about one of the axes, and thus two sets of measurements must be taken. The measurements may be repeated for a few chosen distances and angles thought likely to be useful in the particular applications under consideration.

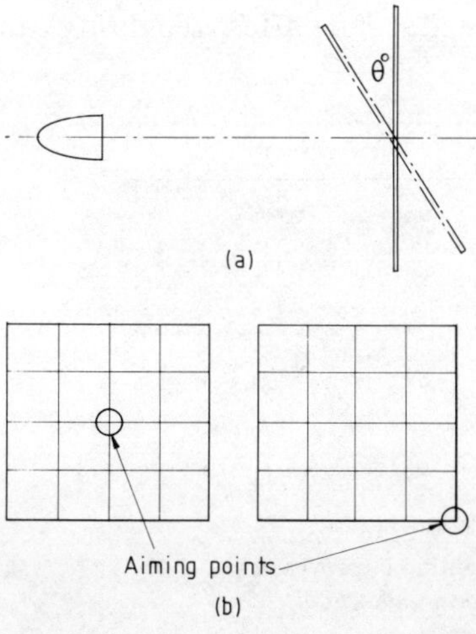


*Figure 17.5* Use of measurement-board to record illuminances in local lighting studies (see also Figure 17.4). (a) Board may be tilted through angle θ to make measurements at any incident angle. (b) For wide beams, it may be more convenient to aim at a corner, noting that if the axis of the beam is not perpendicular to the plane of the board the distribution will not be symmetrical about the aiming-point.

Chapter 18

# Economics of good industrial lighting

A reluctance to invest in good lighting is very often based upon the idea that only investments in production plant can generate profits. But, improved lighting in any industrial premises is likely to make the utilization of all other resources more efficient, and will also reduce the risks of loss due to a variety of causes. This chapter reviews the ways in which calculations may be performed to establish that investment in good lighting is economically sound.

## 18.1 Calculating the cost-benefit of lighting

**18.1.1** Cost-benefits attributable to electric lighting systems may be calculated on a number of bases including savings due to better inspection (Chapter 3), savings due to integration of lighting and heating (9.4), savings due to improvements in output and better quality of work (1.2), savings due to reduction accidents (1.3) and many others. The problem in calculating the cost-benefit of lighting may well be to ensure that all the beneficial factors are taken into account. The effects of provision of better lighting will include matters of productivity and quality, and the health, safety and morale of personnel (Figure 18.1).

**18.1.2** Many studies of cost-benefit of lighting employ case-histories in which the work is initially poorly lighted, and after relighting an

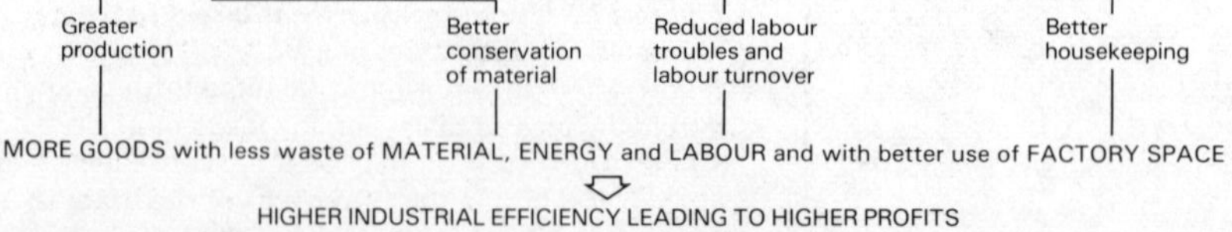


*Figure 18.1* Diagram indicating ways in which better lighting improves profitability. From 'Better Industrial Lighting—The Economic Case'[54]

improvement in profitability is revealed. In most cases the extra cost of the lighting is but a tiny fraction of the extra value of production or its profitability(54) (55). As was long ago pointed out by Prof. L. Schneider, the expected increase in efficiency of working will depend not only on the range of change of illuminances but also on the difficulty of the visual task, i.e. in critical visual tasks even a small degree of under-lighting will severely handicap the work, and correcting this will produce remarkable improvements in output (Table 12), while massive improvements in illuminance cannot more than marginally improve work performance on simple visual tasks(4).

**Table 12** (Paragraph 18.1.2) Improvement in work performance on improvement of illuminance for different grades of visual task difficulty. (Professor L. Schneider, CIE, Vienna, 1963)

| *Illuminance change (lux)* | *Grade or class of work* | *Expected increase in performance (%)* |
|---|---|---|
| From 100/150 to 1000/1500 | Medium | 5–6 |
| | Fine | 9–12.5 |
| | Very fine | 15.5 |
| | Minute | 40–50 |

Only a very small improvement in productivity (say of 0.5%) may be all that is needed to pay for the improved lighting; any further improvement in productivity must lead to improved profitability.

**18.1.3** In the architectural design of buildings, adoption of a suitable daylight factor (2.3.4) incurs costs in the provision of windows, and possibly involves an increase in ceiling heights in order that daylight can reach into the inner portions of large rooms. Theoretical studies for many buildings show that a combination of electric lighting and such daylight as may be available, or electric lighting alone with no provision of windows, is generally of lower total cost over the life of a building than the attempt to provide sufficient daylighting to work by for such hours as it may be available. The hours of availability of daylight are limited, and do not always coincide with the hours of working. Substantial and continuing savings may be made by relying mainly on electric lighting, and regarding any available natural lighting as a bonus. The savings come about by such factors as (a) reduced building cost by not having to provide windows and subsequently maintaining them and cleaning them; (b) possibly reduced building cost by being able to reduce ceiling heights; (c) reduced heating costs because thermal losses through the outside walls are reduced; (d) reduced summer cooling cost because less heat is gained through outside walls in summer. Because the heating and cooling loads can be so reduced, it becomes possible to achieve all-year-round comfort conditions with smaller capital cost for the air-conditioning plant, and lower operating costs (Chapter 9). If natural light is to be admitted to a new building, it can be arranged that the heat losses and gains are minimised by using small windows and double-glazing them, and ensuring that shafts of sunlight cannot enter in any area where critical visual tasks are performed.

With these provisos, natural light can be admitted for its social and human attributes, not for its value as illumination, and the cost of this amenity will be small compared with other benefits to be expected from a well-conceived and executed plan for lighting the building. Various critical visual tasks, especially those involving colour judgment, have hitherto been considered as only capable of performance to satisfactory standard under natural lighting; but this premise is false, and it can be shown that the unvarying spectral composition, quality and quantity of electric lighting is demonstrably better for these critical visual tasks (Chapter 4).

**18.1.4** In the practical work of running almost any kind of factory, there is continuing loss due to minor accidents to goods and materials, resulting in items being spoiled and becoming scrap, or requiring re-working, re-polishing or re-painting etc. Such small deteriorations occur in handling, storing, issuing etc at stores, and in transport of the items in the factory. These multifarious minor incidents occur all through the life of a piece-part or assembly, and thus usually do not become accounted for as a 'cost-centre'. It has long been realised that much of this minor damage to components and materials is because the lighting provided is not sufficient and suitable for their safe and proper handling. In a study carried out at a sheetmetal factory where components were made, painted and assembled, responsible managers were asked to estimate the damage rate and the cost of rejection or rectification of damaged items in manufacture. They were then asked to estimate by what percentage this damage rate might be reduced by the provision of better lighting. It later was found that their estimates on the latter score were rather low, but the overall estimate was that 2 per cent of all components going through production needed rectification for dents, burrs, scratches etc received after being passed through inspection, and that over 10 per cent of all finished paintwork was either stripped and re-painted or required re-touching by hand to correct defects before the final 'passing' of the complete assembled products. A programme of relighting was carried out, bringing the illuminances throughout the works (including the stores areas) up to *CIBS/IES Code*(5) levels. Because of the diffuse nature of the losses, it was not possible to positively quantify the savings in relation to the lighting, but the annual benefit was estimated at 20 per cent of the total annual cost of the improved lighting. Other cost-benefits in productivity and quality were also observed.

**18.1.5** It would be misleading to suggest that where better lighting facilitates greater productivity the entire improvement is to be recorded as greater profit. If output is higher, then doubtless more materials have been consumed, and possibly operatives have been paid greater bonus or piecework earnings too. Calculations of cost-benefit should be adjusted to take account of this.

**18.1.6** It is a sad fact of modern society that improvements in efficiency of production result in a smaller workforce being required to achieve a particular standard of output. The impact of the computer and the microprocessor has not yet been fully felt, but there is no doubt that companies can operate with fewer workers as they improve the means of production and if the output level remains constant. Of course, conversely, as the efficiency is improved, a given labour force can produce greater output which is a great advantage if markets for the greater output can be found. Not unnaturally, there have been examples of workers and Trade Unions resisting new production methods (fearing that they would lead to loss of jobs) and bargaining for higher wages which tend to reduce the nett profitability to the company and thereby reduce the incentive of the company towards modernisation. If there is any tendency to this sort of 'whiplash', with staff resenting the installation of improved lighting because they fear that a resultant increase in productivity will threaten jobs or earnings, happily this seems to be offset by the appreciation of the staff of the comforts and personal benefits better lighting brings. The amenity value, and the value of lighting to enhanced safety, is sufficient to make staff universally pleased to have better lighting at their place of work. However, this fact may be obscured if 'relaxation allowances' are paid to compensate for poor conditions (1.2.2). It will be found difficult to withdraw such allowances when the lighting is eventually brought to satisfactory modern standards, and this should be allowed for in calculations of cost-benefit.

## 18.2 Economic justifications and Pay-back Period

**18.2.1** The proportion of the national energy consumption used for lighting is quite small, probably around 6 per cent, yet lighting is an important factor in the lives of everyone. In typical factories, the part of the energy consumption attributable to lighting may be between 0.5 per cent (in factories with high energy usage) up to around 15 per cent (in labour-intensive industries with relatively low energy usage). Because efficacies of lamps have risen over the years due to research and development, this has offset the rising cost of energy, so that lighting still is relatively cheap. The average annual lighting cost (i.e. the Ten Year Cost divided by 10) (18.2.2) seems to average out at around 0.5 to 1.0 per cent of the annual wages and salaries bill of the staff who will use the lighting. This seems to work out quite consistently; better paid staff tend to need higher lighting levels to aid them in performing their more skilled work. Lighting therefore forms only a minute fraction of the total costs of manufactured goods. Taking a national average of the labour content of all manufactured goods accounting for 40 per cent of the cost, then the fraction of the cost of the goods accounted for by lighting is around 0.2 to 0.4 per cent. There can be few factories which have a costing system so tight (and so accurate) that they could not afford to have the very best in lighting, even ignoring the cost-benefits which accrue in almost every case.

**18.2.2** In the case of new building projects, calculation of the Pay-back Period must be based on a notional comparison with a possible alternative scheme which involves lower capital cost and has the potential to incur higher running costs and/or a lower potential for creating efficient and safe factory conditions. In the case of new lighting going into existing factories, the savings are real, and the benefits capable of careful quantification (18.3). In the case of existing lighting installations over ten years old, in most cases substantial improvement in illuminance, glare control and colour quality can usually be made with a reduction in energy costs and maintenance costs, so that the investment yields a Pay-back Period of between three years and six months. At the end of the Pay-back Period, the cost will have been equated by the savings to date, and thereafter the savings will continue through the life of the installation.

## 18.3 Cost comparisons with inflation adjustment

**18.3.1** There are two economic factors which should enter into studies of cost of capital equipment and its operating cost. The first is that had the money not been invested in the capital equipment, it would have been available for investment in something else, or could have earned money by interest. The second factor is that we have high rates of inflation world-wide, and therefore all forecasts of the cost of operating the equipment should be weighted with an annual inflation percentage. As both interest rates and inflation rates vary from place to place and with time, it is only possible to outline the principle of the calculations. Without going into the theories of *discounted cash flow* and *future accounting* (both of which might need to be considered in the case of large capital sums being invested in lighting), it can be stated that, at the time of writing (when bank rates and inflation rates are of a very high order), it is necessary to include a notional value of the interest lost by investing the money in lighting, and to assume that the costs of lamp replacements, energy, labour etc will also increase during the life of the installation.

**18.3.2** Economic comparisons to assess the worth of proposals involving capital expenditure are usually limited to examining the capital costs and, in most cases, choosing the cheapest of the alternatives offered. Bargains bought this way can turn out to be costly, if the lower capital cost proposal has a high running cost; so it has become customary to prepare a break-even point diagram (Figure 18.2) to compare two or more proposals. Such

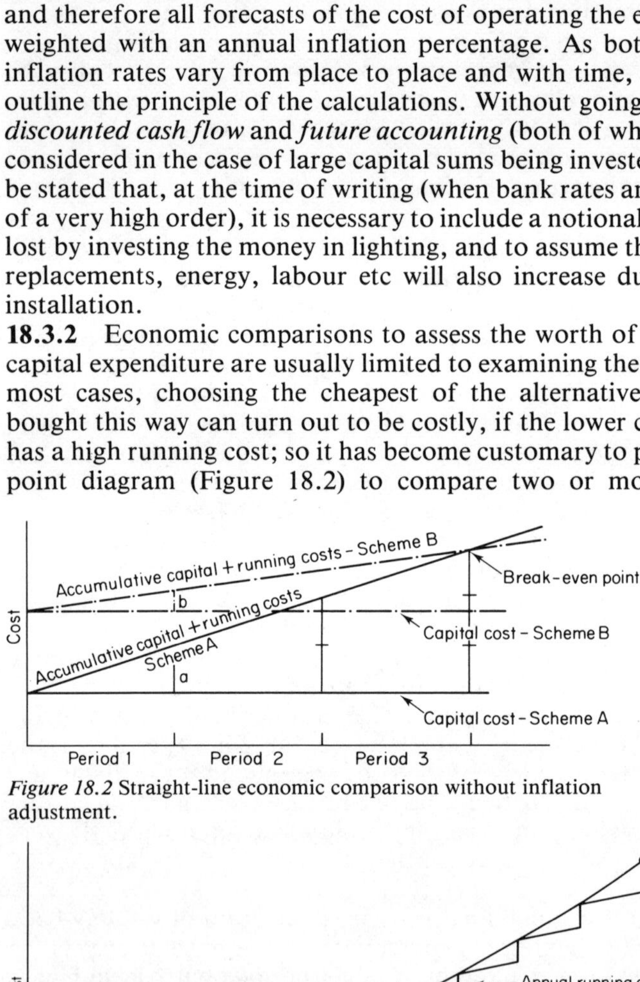


*Figure 18.2* Straight-line economic comparison without inflation adjustment.

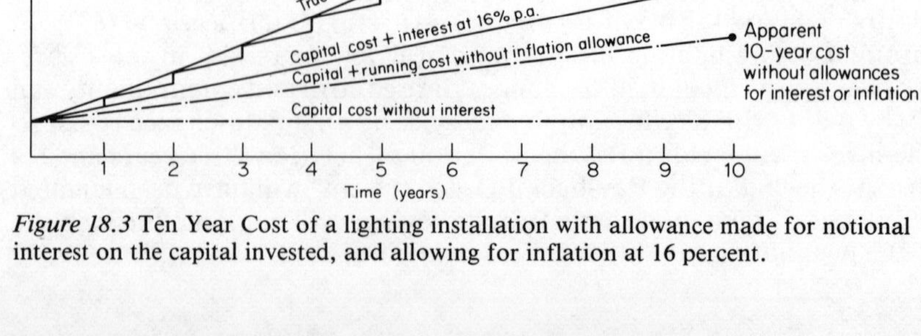


*Figure 18.3* Ten Year Cost of a lighting installation with allowance made for notional interest on the capital invested, and allowing for inflation at 16 percent.

calculations make the false assumption that all costs are fixed, whereas in the real world of business today, costs continue to inflate long after the capital cost is laid out. It would be sensible to assume that the cost of labour, replacement lamps, energy etc will rise during the technical life of the new lighting installation. Taking a round figure of 16 per cent annual inflation as an example, it can be calculated that the true ten-year cost is likely to be around three times the straight-line present value calculation (Figure 18.3). Even the capital sum expended must be subject to an inflation adjustment, on the basis that if the money had not been spent on a new lighting installation it could have been invested in some profitable machine tools, or could have financed trading, or merely been put in the deposit account at interest. This inflation of the capital cost is shown in the diagram. It therefore becomes necessary to calculate the cost of a lighting installation with some care, and to set it against the savings that will justify its outlay (18.2.3). The basic calculation of lighting costs is

Capital cost = Cost of lighting equipment + cost of electrical installation + cost of any permanent access means + notional cost of any floor space unusable because of access requirements (13.1.3) + cost of initial set of lamps.

Annual running cost = Cost of electrical units consumed + proportion of related maximum demand charge + depreciation.

Annual maintenance cost = Annual cost of replacement lamps and consumable items + annual cost of cleaning and relamping + annual cost of preventative and corrective maintenance (materials + labour) + incidental costs (e.g. hire of access equipment etc).

These factors combine thus

$$\text{Ten Year Cost} = (C+R+M)K$$

where $C$ = Capital cost

$R$ = Annual running cost

$M$ = Annual maintenance cost

$K$ = Assumed rate of annual inflation, per cent, over 10 years at compound interest.

Similarly,

$$\text{Annual Cost} = \frac{\text{Ten year Cost}}{10}$$

**18.3.3** The Annual Benefit from the new lighting will comprise components such as

Savings on capital cost of building (18.1.3);
Savings on building energy (Chapter 9);
Improved added value to manufactures due to improved quality;
Value of extra production (less direct manufacturing costs of materials and labour);
Savings in running costs and maintenance costs as compared with less efficient system;
Savings due to reduction in damage, scrapping, rectification;

Savings due to better staff relations, reduced labour turnover leading to lowered costs of recruitment and training;
Savings due to reduced absenteeism (physical absences including latecoming and early departure) and invisible absenteeism (where employee is present but not actually working);
Savings due to reduced accident rate (lost time, management time, claims against the company), and in the minor sickness rate.

The monetary values assigned to the foregoing components and any others which apply should be added to find the Annual Benefit, and then to calculate the Pay-back Period. For example, if the Annual Benefit from the lighting was estimated at £5,000, and the Annual Cost of the new lighting was £10 000, then the Pay-back Period would be 24 months. However, the Annual Benefit should be discounted by the Annual Benefit of the previous lighting, and the Capital Cost of the new lighting should be partly offset by the scrap value, if any, of the old system. This then gives the Nett Annual Benefit, which will continue through the life of the installation, say at least 10 years (18.4).

## 18.4 Tax allowances on lighting investment in the UK

**18.4.1** Taxation allowances for business investment vary from budget to budget, and it is therefore impossible to set out the details. But, the general principle followed in the UK is that investment in any capital plant which is intended to improve productivity or protect against loss (including expenses necessary to comply with law's requirements for safety and health) are regarded as earning some relief from taxation on profits. The relief is allotted to the tax-paying organization in stages, part in the first year, and a tapering allowance in subsequent years. It should be noted that this relief is set against profits, and does not (at present) come in the form of a grant. In effect, the Government, by foregoing part of the tax that would otherwise be payable, contributes to the cost of the capital investment.

**18.4.2** There is usually no problem in qualifying investment in better industrial lighting for tax relief, but it may happen that the inspector will need to be satisfied that the lighting is of such nature that it is required (a) to comply with legislation on health and safety (see Appendix I); (b) to enable the processes of manufacture, inspection, store-keeping etc to be done efficiently, or (c) to protect against loss which might occur if the lighting was not provided. The types of lighting described in this book would almost certainly qualify for tax relief in every case; but it could be necessary to justify lighting which had the appearance of a luxury or unnecessary expenditure. For example, building floodlighting might be installed which contributed significantly to the night security of the premises (15.3), and the Inspector might require an explanation of its function. When the taxation relief on the capital investment in the new lighting is known, the calculation of the Nett Annual Benefit may be adjusted accordingly.

Chapter 19

# Examples of lighting practice in industries

Environmental conditions in the various industries differ so greatly that it is impossible to devise a standard lighting method that will suit all industrial areas. Much that is written in this book is of general application, but in applying the principles outlined in particular factories one must take account of the specific visual tasks to be performed in each, and adapt the lighting to the kinds of buildings met, choosing the lighting equipment to suit the hazards and environmental attacks that it must withstand if it is to give safe and durable service. The brief survey of lighting in a number of industries given in this Chapter cannot claim to be comprehensive; but the examples chosen will indicate the range of problems likely to be met in practice; further, ideas which have been successfully used in one industrial application may often be utilized in others having some common visual problem or environmental condition.

## 19.1 Food, drink and pharmaceutical industries

**19.1.1** *Food industries.* The lighting requirements for these industries are dominated by two important factors: the need for lighting of good colour-rendering (4.2.3) to enable foodstuffs to be inspected for freshness, variety and quality; and the requirement that the lighting installation shall comply in all respects with the requirements of the Food Hygiene (General) Regulations[48]. Dealing first with the problem of colour-rendering, it should be noted that many foodstuffs have unexpected optical properties, such as exhibiting a degree of fluorescence under daylight and u.v.-rich sources, or demonstrating *dichromaticity* in the reflected or transmitted mode (4.3.10). The general case is that foodstuffs are sold in retail outlets under uncontrolled conditions of lighting; there may be admixtures of natural light with that from virtually any kind of lamp. One situation which may be to the disadvantage of the food manufacturer is where food in the factory is inspected under, say, *Northlight* or *Colour Matching* tubes, but is shown for sale under *De luxe Natural* tubes. It can only be advised that, in order to ensure that the foodstuffs look attractive at the point of sale, the manufacturer should inspect under a high colour quality source for freshness, variety and quality; and that he should also view his products under such tubes as *Warm White, De Luxe Natural* and *Natural* tubes from time to time to ensure that the colours of products such as pastry, confectionery etc look attractive under those sources too. It should be noted that butcher's meat in particular should be carefully inspected under a good colour-rendering cool source for quality, for meat almost at the point of putrifaction will appear passably good under *De Luxe Natural* tubes or under *Warm White* tubes sleeved with pink clear plastic. (Meat and fish at or below freezing point have virtually no smell whatever their actual condition.)

**19.1.2** Compliance with the very reasonable objectives of the *Food* Hygiene (General) Regulations[48] is mainly a matter of using only luminaires that will satisfy the requirements. The Regulations require that there shall be no likelihood of any part of a luminaire or lamp falling into foodstuffs; thus, any openable parts should be hinged or connected with chains etc, and all fixings should be captive. The lamp must be enclosed so that, if it should shatter, broken parts cannot fall into the food area. As regards the risks of contamination of foodstuffs from dust and dirt on or in luminaires, the luminaires should present the minimum horizontal surface upon which such may settle (Figure 19.1).

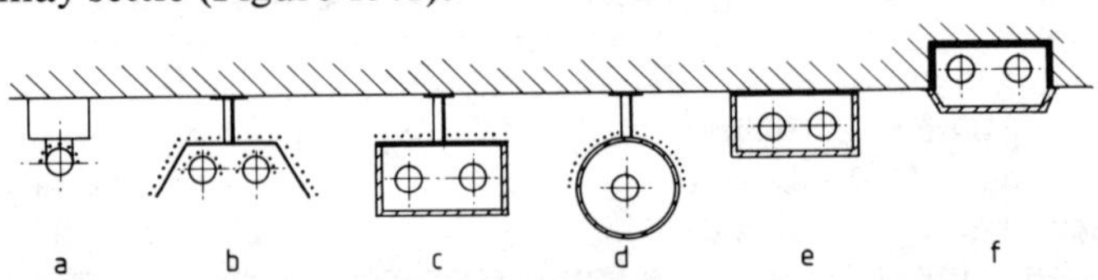


*Figure 19.1* The six most commonly used of fluorescent-tube luminaires, in ascending order of hygiene preference from left to right. The dotted surfaces are those upon which dirt may settle and later fall to contaminate surfaces and foods. Luminaires types *a* and *b* do not comply with the Regulations unless the tubes are enclosed in special transparent tubular enclosures, eg as in waterproof batten construction.

**19.1.3** Luminaires which are enclosed but which are not dust-tight present the danger that dust falling into the luminaire may later become entrained in an air-current and fall to contaminate surfaces or foods. Only luminaires which are dust-tight or dust-proof to BS 4533[27] will satisfy the hygiene requirements in this respect, and even these are not free of the dust problem if they are suspended so that dust can settle on their upper surfaces.

**19.1.4** The need for better colour-rendering tubes in food preparation has been mentioned (19.1.1), and it would be sound practice to use tubes such as *Northlight* or *Colour Matching* where meat, fish or poultry is handled; but for other sections of the food industry, there may be an element of personal choice; all the tubes in the upper half of the list given in Table 5 (para. 4.3.5) are used and found satisfactory in the baking and confectionery trades. The lamps are listed in decending order of colour temperature, not in their order according to their colour-rendering properties. For larger food establishments with higher roofs, Mercury Halide (MBI) lamps or the fluorescent-bulb version of these lamps (MBIF) may be used successfully for all manufacturing tasks.

**19.1.5** When premises have been relighted to modern standards, very often the management immediately have the interior redecorated, perhaps because when there is enough light they can see just how grubby and dingy the place has become! Personal hygiene habits seem to improve, too, when it is easy to see that fingernails and whites are not spotless. There is no doubt that the provision of adequate illuminance is an important factor in promoting high standards of hygiene in the food, drink and pharmaceutical industries.

**19.1.6** In canning lines at food factories, there are often overhangs of equipment which will cause shadowing, necessitating the placing of local lighting units within the machine lines. These should be robustly constructed or protected against accidental damage, and each work position should be carefully checked to ensure that excessive glare is not caused to operators from these additional luminaires, and screening of some kind devised if necessary. Commonly, enclosed batten luminaries, water jet-proof to BS 4533,[27] are employed for these topping-up duties. The effect of steam from autoclaves etc, and the effect of heat from adjacent steam lines should be considered to prevent early deterioration of the lighting equipment and wiring.

**19.1.7** *Drink industries.* Under this convenient title, the lighting needs of the milk bottling (19.1.8), soft drinks industry (19.1.9) and brewing and distilling (19.1.10) may be briefly reviewed, mentioning some of the points peculiar to these activities in respect of the visual tasks and environments that may affect the methods of lighting.

**19.1.8** *Milk bottling.* The visual problems involved in the inspection of glass bottles in single-line progression past an inspection station are discussed under Glass industry (19.4.3). Apart from the inspection stations, and the local laboratory (which should be lighted as for any other laboratory, see Pharmaceutical industry (19.1.3)) all areas at a milk bottling plant, interior and exterior, may be efficiently lighted using high pressure sodium (SON) lamps. The luminaires in the interiors should be water jet-proof pattern to BS 4533[27].

**19.1.9** *Soft drinks industry.* The general requirements are as for milk bottling (19.2.2), noting that canning lines may require special care (19.1.6).

**19.1.10** *Brewing, distilling.* It could be a sound move to standardize on a corrosion-resistant Zone 2 luminaire for use many locations in distilleries and breweries where wet, dusty, and sometimes flammable conditions may exist. The bottling areas will be similar in general problems of lighting to milk bottling (19.2.2), and the canning areas have the same lighting problems as other canning lines (19.1.6). There can be a preference in breweries for luminaires to be hook-suspended and connected by plugs to local socket-outlets, so that luminaires can be removed complete for cleaning and servicing in the workshop. A few spare luminaires enable them to be substituted at once, avoiding any lengthy outage and completely preventing contamination from dusts which may contain fungii and yeasts which the brewer must prevent from causing contamination. Both in distilling and brewing, the nature of the plant is that there is no uniform 'working plane', but tasks are performed at various levels throughout each building. The lighting must be adapted to these conditions, placing local lights where gauges are read, ensuring that walkways and ladderways are safely lighted, and studying each working position to ensure that topping-up lights do not cause excessive glare. For periodic access to high plant for cleaning and maintenance, it is a sound idea to position a number of tungsten-halogen floodlight luminaires in the upper reaches of each high plant room; these will only be switched on when work at high level is to be performed. The TH lamps will withstand plant vibration better than will GLS lamps; having a nominal life of 2000 h, and being used only for short intervals over a long period, they may give several years of service before needing replacement. The fact that alcohol fumes are highly flammable and explosive is sometimes overlooked. Consultation between client and lighting engineer should resolve if there are danger areas which will require protected luminaires. Nor should it be forgotten that finely particulate dusts, such as grain dusts, if suspended in air at the right concentration, can burn and explode with violence (11.1.9).

**19.1.11** *Pharmaceutical industries.* The hygiene requirements in the pharmaceutical industries are similar to but more stringent than those for the food industries generally, though the basic principles outlined so far in this chapter may all apply. The special requirements relate to sterile rooms and clean rooms which have been described (10.3). For the packing of drugs, dressings, sutures and instruments etc, cowled benches with individual lighting and local forced-draft ventilation may be employed. For these, the examples quoted for bench-top inspection booths (3.2.7, 4.4.10) may be useful guidelines. In some laboratories merely switching out some of the luminaires may not give sufficient control when a diminished illuminance is required, the employment of solid-state dimmer circuits on some local or general lighting luminaires may be needed. Some laboratories will require Zone 2 or Zone 1 luminaires if work which produces a flame-hazard is done. Note that if a laboratory is declared a hazardous Zone, all the electrical equipment (including bench lights and socket-outlets etc) will have to comply with the electrical and temperature constraints of that Zone (Chapter 11). In areas where high standards of cleanliness or sterility are required, and particularly if this is accompanied by a hazard due to atmospheric contamination with dusts or flammable substances, the advantages of using pressurized luminaires and electrical systems should be examined (11.1.10). The visual tasks in clean and sterile areas, including the critical tasks at workstations in these areas, may require special lighting and ventilating arrangements (10.3.2).

## 19.2 Clothing, textiles, paper and leather industries

**19.2.1** *Clothing industry.* An extensive study by the author of the lighting conditions in the clothing industry indicated that this industry is highly sensitive to improvements to the working environment, and that lighting was a factor of high importance in creating conditions of high productivity and staff contentment. Studies at a number of clothing factories were carried out under the auspices of The Electricity Council and the Clothing Institute, and some early results were published[49]. One important fact noted was that some 60 per cent of all faults found in garments at final inspection were fabric faults, i.e. had there been better inspection of the cloth initially, considerable costs in marking out, cutting, machining, finishing etc to produce an unsaleable or substandard garment could have been avoided. It is noted that lighting at cutting tables is rarely up to standard in quantity or colour quality to enable the cutter to spot fabric faults. And it should be asked if the cutting table is the place for this inspection; would it not always be better to pass the fabric over a winding frame and have an examiner to check it properly? If this is done, weave faults can be revealed by back-lighting, or by placing light on the fabric at a glancing angle. It is apparent that an important quality gain could be made this way, with substantial savings (3.2.4, 19.2.2).

In an industry that employs so many females, the effect on morale of providing a pleasant environment is important; the lighting aids the work as well as accounting for most of the appearance of the workplace as seen by the employee at the workbench or machine. The operatives suffer a sense of isolation due to noise which prevents conversation, and much of the work, though simple and repetitive, requires constant attention. It is clear that lighting helps the maintenance of vigilance in such situations, aiding not only vision but mental concentration[50]. Lighting can help offset the feelings of dissatisfaction many young female workers feel with the working environment, making their jobs more satisfying, and helping the employer to reduce the costs of recruitment and training of new employees. This statement is particularly true in the clothing industry.

**19.2.2** *Textiles.* In the manufacture of both natural and manmade fibres, the problem of dust affecting the luminaires must be dealt with. It is not uncommon to see luminaires festooned with dust, which can so insulate a luminaire as to make it overheat; there is thus some degree of fire risk. The repetitive nature of the work, particularly in machine-watching, carries with it a degree of boredom; so that if workers are to be vigilant and of good morale, the lighting must be adequate and help to create a cheerful workplace. When the lighting is such as to enable the smallest weaving faults to be seen, it is sometimes found that there are more stoppages for fault correction; but the gross stoppage time will be less, with a considerable saving in menders' time at a later stage. Because the machines require adequate lighting on their vertical faces, regular patterns of overhead luminaires may not suit many spinning and weaving shops, and luminaires must be localized to put the light where it is needed. Machines, especially those parts against which the thread is seen, should be painted in matt pastel colours to give good contrast with the yarn. For final grey perching, an 'artificial window' may be used; this consists of a wall-mounted or free-standing frame containing an array of fluorescent tubes, either bare or behind a diffusing cover, perhaps as many as 12 or 15 65W tubes spaced closely may be used. The artificial window is placed parallel to the vertical cloth to be inspected, and about 1.5 m from it; the operator stands with his back to the array of tubes, which is of large enough area to throw only an almost imperceptible shadow on to the cloth from the inspector's body (Figure 19.2). Some fabrics are better inspected by rear-lighting, the fluorescent tubes being placed in a glazed frame which can be tilted and varied in height to suit the operator's needs (Figure 19.3). It is desirable that the tubes in both artificial windows and rear-lighting frames be equipped with dimmers, so the operator can vary the tube output to suit his task. One equipment can serve the purposes of both acting as an artificial window and a rear-lighting unit, and will do so the more conveniently if fitted with castors or wheels so the unit can be readily moved and the castors or wheels braked when required. These lighting units are excellent applications for *reflectorised* fluorescent tubes (Appendix III) though it may be difficult to obtain these with high colour quality phosphors—apparently this combination of features is not sufficiently frequently asked for to justify regular stocking by suppliers.

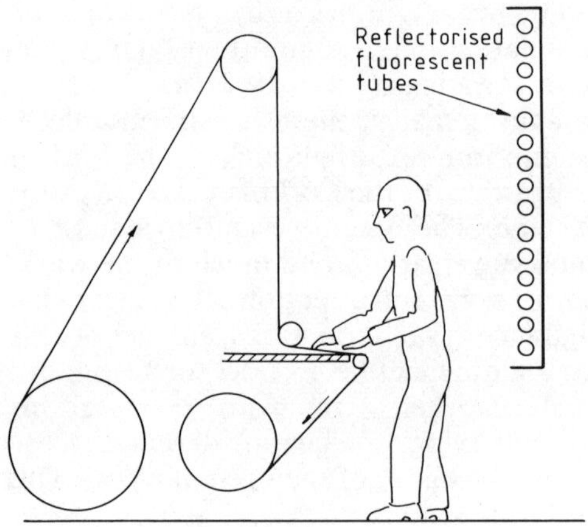


*Figure 19.2* Use of an 'artificial window' for perching (inspection) of cloth.

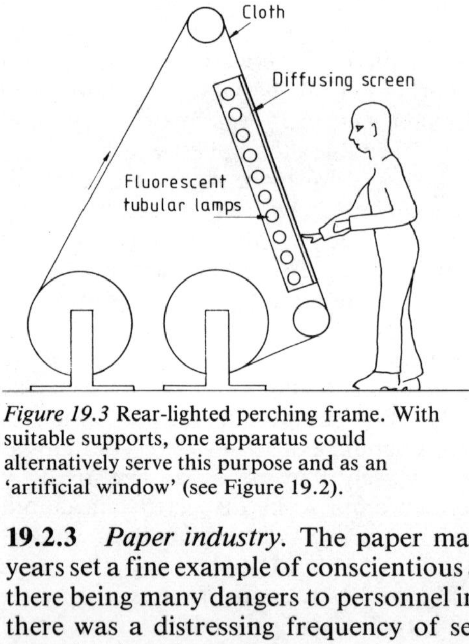


*Figure 19.3* Rear-lighted perching frame. With suitable supports, one apparatus could alternatively serve this purpose and as an 'artificial window' (see Figure 19.2).

**19.2.3** *Paper industry.* The paper manufacturing industry has for many years set a fine example of conscientious application of good safety practices, there being many dangers to personnel in this industry. A decade or so back, there was a distressing frequency of serious accidents at the reel-up, and investigation showed that the risks were greater when the lighting in such potentially dangerous situations was even marginally inadequate. The provision of adequate illuminance encourages operators to use their eyes, not their fingers, to ascertain if paper is winding tight and travelling true, so that the risks of accident are much reduced. The paper industry employs only small numbers of persons in relation to the value of its turnover, as the industry is highly mechanized. Lighting is mainly required for general supervision of machines; but for setting-up, lacing machines etc, there must be light projected into the sides of the machines. For this, angled luminaires are sometimes effectively employed. These are particularly helpful for urgent breakdown work, and for clearing jams etc, the essence of obtaining profitability being the ability to keep the machines running with minimum outage time. Back-lighting may be used to watch the web of paper for defects, i.e. luminaires may be placed within the machine behind the web as seen by the minder. Enclosed luminaires are favoured for use in machine halls, where there is a damp atmosphere. In the works laboratory or test room, lighting should be of a standard similar for general laboratories (19.1.11) with special attention to colour (Chapter 4). It should be noted that judgment of colour of wet pulp is a constant difficulty where precise control of colour is required, and the rules for good colour work have to be followed (4.3, 4.4). In paper mills where materials are stacked out of doors, drivers of trucks, clasp-trucks etc may be frequently passing between the indoor and outdoor areas, so that some zones of intermediate illuminance may be needed (2.3.2). See papagraph 3.2.6d regarding lighting in salles.

**19.2.4** *Leather industries.* Although there is these days less working of natural leather, the demand for quality is higher than ever before. In the treatment, dyeing, grading, cutting etc of leathers, the provision of excellent lighting is a first requirement for good work, often with a need for directional lighting (3.2.6d). In the boot and shoe industries, a number of classic studies have clearly demonstrated the relationship between good lighting and high productivity coupled with low accident frequency, e.g. the Mossbach Gruber case history[(4)]. By the very nature of shoemaking machinery, particularly in clicking and closing, it is impossible to guard the machines; adequate illuminance is a valuable safeguard for fingers under these circumstances. In some factories, optical guidelines are projected on to the work, and the lighting must be carefully devised to provide sufficient lighting without swamping these so that they cannot be clearly seen. Final inspection stages for all leather goods and shoes justify a separate inspection department, with the provision of high quality lighting. In particular, for the final inspection of highly glossy leathers, patent leathers etc, the use of a large-area fairly low brightness luminaire over the inspection station is essential.

**19.2.4** In the group of industries covered in this section (clothing, textiles, paper and leather) many of the products consist of organic basic materials which have been treated with dyes to produce the final product. The organic bases, plus the organic or inorganic colourants used, may result in products which undergo significant change in apparent colour according to the nature of the illuminant. Natural light, which contains a small proportion of ultraviolet light, may cause these substances to fluoresce, so that the colour seems to change. The manufacturer may need to inspect both raw materials and final products under several alternative illuminants to satisfy himself that the product will be satisfactory. To quote just one example of this colour change effect, a clothing manufacturer had a large quantity of skirts returned because some panels in the skirts seemed to be of a markedly different colour from the others in daylight, though all skirt panels seemed to be identical in normal electric lighting; it was only when the cloths were checked under Northlight tubes that the colour difference (due to slight variation in dyeing between rolls of cloth from the same supplier) became apparent. More information on this important aspect of lighting for manufacture is given in Chapter 4.

## 19.3 Engineering, plastics, printing and furniture trades

**19.3.1** *Engineering.* In the production processes of metal-removal, metal-deformation etc, although there is a high degree of reliance on machine processing, the man/machine cybernetic relationship is dominated by the visual sense; thus, in these industries, the provision of good lighting is a first prerequisite for productivity as well as for the safety and well-being of the operators. In many engineering production shops, better use could be made of the human and material resources by the provision of an environment of cheerful appearance (4.1). Here is an extract from a report on a machine shop in the vehicle industry: "In large parts of the factory, the general decor of the building and machinery is green, and the floors are near-black. The result of this combination is to produce a very depressing effect indeed, which cannot be good for morale. In some departments the lanes between machines are very narrow, and machinery extends to above eye-level, which, with the dark colours and poor lighting must produce a depressing, even hysterical, response in sensitive workers. One must ask if such conditions are conducive to good mental health and the welfare of the workers, let alone their productivity." Sadly, such conditions are not exceptional, and thus opportunities to improve productivity and provide a humane and comfortable environment for the employees are lost through lack of understanding on the part of managements. It saddens the author to remember that he has been trying to bring about change in engineering production shops for over thirty five years, and still there are 'dark Satanic mills' which could be transformed with quite minimal expenditure and with good cost-justification for that expenditure (18.2). In machine shops, the benefits of inter-process inspection on the shop floor may be very marked (3.2.4).

In microtechnology processes as used in the computer industry, it is to be noted that too great a reliance on small-area task-lighting and built-in lights within microscopes and micromanipulators etc may be counter productive. In one department assembling minute components, the girls were initially provided with general lighting at 350 lux, and had built-in lighting within the optical devices they worked at, giving about 1500 lux on the point of work as viewed. The girls suffered fatigue and headaches, and the management found that it was necessary to institute a routine of 30 minutes work followed by 10 minutes of eye relaxation. This limited the productivity of the unit, but it was found that any longer work periods or shorter rest periods reduced the quality of the output. After the tasks had been studied, the room was experimentally relighted to 1750 lux of general lighting, and some of the workstations were supplied with small focusing spotlamps using 35 W motor headlamp bulbs on a 12 V supply controlled by individual dimmers. After an experimental period, it was found that work periods of 1 hour were now possible without loss of quality or complaints of headaches or eye-strain. Not all the operators took to the use of the external lights at the workstations; some preferred to use the built-in lights in the magnifiers either alone or additionally to the spotlights. It was generally agreed that the fact that the spotlights could be set to direct light to the task at a flat angle was an advantage. The opinion expressed by the girls was that under the new lighting conditions they did not feel so isolated, as they could look away from the microscope for a few seconds without any difficulty of re-adaptation on returning to the task. It seems that the few seconds glancing away from the task without a major change in luminance viewed was very helpful to concentration as it rested the eyes. Incidentally, it was found beneficial to replace the white plastic benchtops with a medium grey matt linoleum (reflection factor 0.4) when the improved general lighting was installed, as the glare-back otherwise was excessive. The walls of the room were repainted a pastel colour of 0.7 reflection factor.

**19.3.2** *Plastics industries.* The manufacture of plastic base materials is carried on in environments indistinguishable from those met in the pharmaceutical industry (19.1.1) and the petrochemical industries (19.4.4), but some notes on lighting requirements in plastics moulding and extrusion shops are merited. In these applications, the presence of solvents in the atmosphere can create a flame hazard for which protected electrical equipment (12.2) will be required. Cases are also known of solvent materials (not necessarily those carrying a high fire risk) affecting the insulation of electrical equipment, or attacking plastic enclosures of luminaires. A common problem in moulding shops is how to get light into moulds so the operator can check that they are free of flash; the top of the machine and the upper mould section mask the mould interior from the overhead lighting. This may be overcome by the use of 'remote local lighting' (2.2.7), perhaps with the use of a metal mirror on the machine to direct light into the mould (3.3.5a). In moulding, pressing and extrusion shops, the air temperature above machine level may be higher than is suitable for the lighting equipment; luminaires in such areas should not be mounted higher than the highest window opening, and forced-draft ventilation may be needed.

**19.3.3** *Printing industry.* Because of the very high visual content in all printing operations, this industry stands to benefit greatly from the application of modern lighting techniques. In the technological transformation of the printing industry that has occurred in the past two decades, many printing concerns have moved to modern premises, often more compact and with better facilities, so that the general standards of lighting are quite good. There are few sections of this book that cannot be construed as having some particular reference to the needs of the printing industry, and it would be tedious to recapitulate all the relevant points. However, it is worth stressing that in printing works where fine colour work is done, there may be some justification in employing illuminances considerably above the current recommendations of the *CIBS/IES Code*[5]; for example, a general lighting system producing 2000 lux from better colour-rendering fluorescent tubes could help to make a litho colour department highly efficient.

Many printworks now employ ultraviolet light for accelerating the 'drying' of special inks; of course, in this process little or no solvent is flashed off, the 'drying' being due to polymerization of the ink molecules. Although uv ink drying lies outside the scope of this book, it is worth noting here that the special lamps used for this purpose must not be placed in ordinary luminaires. Persons working in places where these uv tubes are used should avoid exposure of eyes or skin to the uv radiation; it would, for example, be dangerous to dismantle such equipment and operate it open, or to attempt to by-bass the interlocks on the covers of the equipment.

Another use of non-white light in printing is the employment of a cyan filter in a local task light to enable yellow registration marks to be seen better in litho multicolour printing.

**19.3.4** *Furniture industries.* In the woodworking areas, the problem of dust has to be dealt with; for the penetration of fine wood flour into enclosed luminaires which are not adequately sealed can considerably reduce their light output. In polishing shops, there is certainly a special need for lamps of better colour-rendering, often to be housed in luminaires for Zone 2 risks (12.1). In departments using routing and moulding spindles, circular and band saws etc, there is a possible danger from stroboscopic effect which must be suitably counteracted (5.3.4, 5.3.7). In the saw-doctor's shop, use may be made of a backlit screen for checking saw tooth conformation. Throughout the works there should be consideration of the choice of lightsources; for general machine shops and mills, assembly, stores etc, high pressure sodium lamps (SON lamps) should be the first choice on grounds of efficacy, while fluorescent tubes of high colour quality should be used in veneer preparation and matching, finishing of upholstered goods, polished etc. For veneer sorting, a louvred ceiling may enable the provision of high illuminance with very good control of glare. Local lighting, often built into the machine, may greatly ease the control quality, e.g. by fitting dust-tight batten luminaires at the rear of belt-sanders. Checking of flatness of veneers after pressing may be facilitated by the use of an illuminated grid (3.3.9a) or by an arrangement that seems peculiar to the furniture industry, an apparatus to project light in a controlled directional manner that will aid the detection of flaws (Figure 19.4).

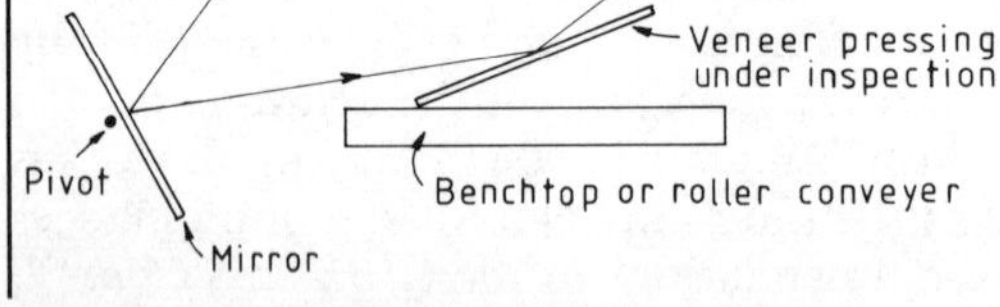


*Figure 19.4* Directional lighting apparatus for examining veneer pressings for flatness. The linear lamps may be reflectorized fluorescent tubes, or double-ended 'architectural' clear tubular filament lamps. The mirror is pivoted to enable it to be set at a convenient angle for the inspector.

## 19.4 Metal industries, foundries, glass, petrochemicals

**19.4.1** *The steel and base metals industries.* Although they use vast amounts of fuels, they cannot afford to waste energy any more than any other kind of business. For this reason, the high pressure sodium (SON) lamp has been taken up and extensively used with great benefit in these industries.

Typical conversions of old fluorescent or mercury-vapour (MBF) installations to SON yield a break-even point in less than two years, sometimes only months; and this may be combined with a substantial increase in illuminance. However, there is still much to be done, and the relighting in these industries is far from complete. Indeed, there are still huge areas that are lighted with tungsten-filament lamps. Except for very temporary installations, the use of these lamps cannot be justified if they are in use for more than say 6 to 10 hours per week. The high-roofed buildings common in these heavy industries causes lamp replacements to be costly in terms of labour; thus replacement every 1000 h or so is uneconomic, and the long life of HID lamps ensures that considerable economy is affected. The general use of gantry cranes results in considerable vibration in building structures, and this can affect lamp life (11.4). Because of the floor obstructions and heat at high level, relamping might be more conveniently peformed through-the-roof (12.3.2) or by raising-and-lowering gear (13.2.12). The drop-forging industry has had a poor record for provision of lighting in past decades, but with the more general use of electric billet-heating, the cleaner atmosphere in the forge does enable better environmental conditions to be maintained. In these drop hammer shops, where the possibilities for personal accidents abound, the provision of adequate lighting is a high priority item for safety. In all industries where hot metal is worked, the general lighting must be sufficient to 'swamp' the brightness of billets, bars coming from the dies etc, or the 'adventitious light' (5.2) will be a hazard due to uncontrolled glare. It has been established by the Technical Committee of the National Federation of Drop Forgers and Stampers that the idea of 'keeping the general lighting level low so that the temperature of a billet may be judged by its brightness' is technical unsound and unjustified on grounds of safety; billet temperature is readily determined by instrumentation.

**19.4.2** *Foundries.* Much that has been written about the steel and base metals industries (19.4.1) is applicable to foundries with the observation that both good morale of the workforce and their safety will depend to a great extent on the ability of the management to ensure that the lighting installation is kept clean in the face of the inevitable pollution that occurs when solid fuels are used. Further, where colour-coded moulding sands are used, the light-source chosen in the moulding department should have sufficient colour quality to enable the sand colours to be identified; only experiment will show if SON lamps are suitable in a particular establishment. However, for all other departments of the foundry (apart from any works laboratory) SON lighting will be effective and economic.

**19.4.3** *Glass industry.* The glass industry contains within itself a number of operations which have quite different lighting needs. While the general case in glassworks in unspecialized, special attention should be given to glassware decorating and to all inspection lines for containers, utensils and drawn cylinders. As regards the inspection of containers (bottles, jars etc) at single-line conveyor inspection stations, fundamental work on the design of these stations has been carried out by the British Glass Industry Research Association in conjunction with manufacturers and The Electricity Council. The essence of a station for inspection at speeds between 50 and 90 pieces per minute is to take advantage of the peripheral vision of the inspector, and to provide a station in which the approaching and receding containers can be

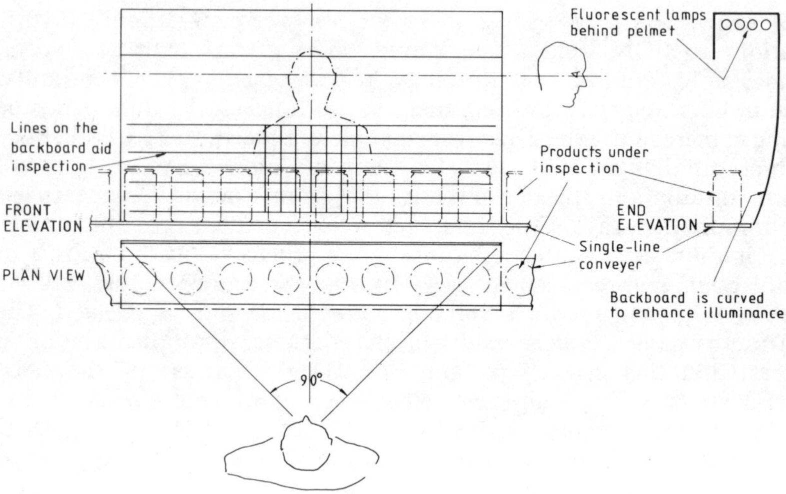


*Figure 19.5* Inspection station for inspecting glass containers on a single-line conveyor.

seen while ostensibly the inspector is viewing the container passing the nominal inspection point (Figure 19.5). This design of inspection station utilizes the idea of an illuminated backboard with a black line grid on it (3.3.9b). It seems that after a period of practice, the inspector carries in his mind's eye a picture of the perfect bottle, and, as though by instinct, will seize and cast into the cullet box any container with airlines, crizzles, hammocks etc, as well as those that are obviously defective at the finish or asymmetrical. The grid device enables him to spot containers with thin or thick walls etc. Tests have shown that the area of the inspection station must be well lighted and very free from direct glare; the approaching and receding conveyer should have an illuminance of at least 500 lux and the backboard needs an illuminance of around 750/1000 lux from the reflectorized fluorescent tubes concealed behind the canopy. The extent of the backboard should be such as

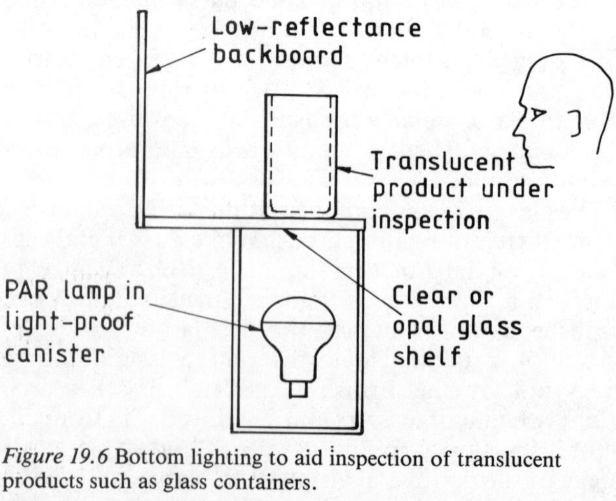


*Figure 19.6* Bottom lighting to aid inspection of translucent products such as glass containers.

to subtend an angle of 90° to the inspector's eyes. Utensils can be inspected very well by bottom lighting (Figure 19.6), viewing the test pieces against a fairly dark background in subdued general lighting. Various means of introducing light into the edges of sheets of glass are used so that the light is 'piped' through the sheet (3.3.7) to reveal faults and surface scratches. Glass tubes and cylinders can be similarly inspected, either by piping the light in from an end, or by placing the cylinder into contact with a prism of clear plastic with a light-source behind it[11].

**19.4.4** *Petrochemicals industry.* In order to minimize hazards from fumes and toxic gases, petrochemical plants tend these days to be built as open-air structures, and the lighting techniques that are needed are mainly those applicable to exterior lighting[1]. Attention is directed in the present book to Chapters 10, 11 and 13, which are particularly relevant to this industry's working environments, and to the general requirements for the lighting of laboratories (19.1.11) where there may be a special need for good colour discernment for chemical tests (4.3, 4.4).

Chapter 20

# Lighting practice in non-manufacturing areas

This chapter deals with the lighting for the non-manufacturing areas of industrial premises, which like every other part of the organization, must make their contribution to the efficiency of the organization. Office lighting, and that for the entrance and circulation areas, may be required to present a pleasant and cheerful appearance, and by this will help visitors form a favourable impression. Lighting in the drawing-office can greatly facilitate efficient work, while that in the staff rooms and amenity areas can help shape the environment of the workers in a favourable way.

## 20.1 Lighting for offices

**20.1.1** Conventional office areas with ceiling heights around 2.5 m to 3.5 m are conveniently illuminated with enclosed prismatic luminaires containing one or two fluorescent tubes. Tube lengths of 1200 mm (4 ft), 1500 mm (5 ft) or 1800 mm (6 ft) may be employed according to the shapes and sizes of the rooms. General purpose fluorescent tubes of WHITE, Warm White, Cool White (Daylight) or NATURAL colours are commonly employed for office lighting, and advantage may be taken of the newer 25 mm (1 inch) diameter range of high-efficacy tubes which are generally available in lengths of 900 mm (3 ft) and 1500 mm (5 ft); a 450 mm (18 inch) tube is also available, and is useful for corridors, toilets etc. Although most office work is not particularly visually demanding, control of glare is important because the workers remain in more-or-less fixed positions through most of the working day. A glare index of 19 is recommended for ordinary offices, with an index of 16 for the drawing office, executive offices and conference rooms. Larger offices may be able to use fluorescent-coated mercury-halide (MBIF) lamps which are now being used in such applications, but this possibility is limited by the available ceiling-height which in turn limits the power of HID lamp that can be used because of its brightness.
**20.1.2** Most offices are lighted by a general lighting scheme which illuminates the whole space uniformly, but, if there are large circulation areas in the space, the luminaires may be localized to the working area. Individual workstation lighting is occasionally used, but its disadvantages (glare to other workers, problems of floor sockets etc) are hardly worth overcoming unless there are special reasons. Control of lighting in offices is important, switching usually being by blocks or rows; in some cases, luminaires over workstations might be controlled with local pull-switches (2.2).

**20.1.3** For larger office areas, particularly those designed on open-plan lines, air-conditioning will probably be employed, and the use of air-handling luminaires should be considered. The advantages of these are (a) the air in the room is extracted through the luminaires, tending to keep the tubes cool and giving enhanced light output; (b) because there may as a result be fewer ceiling grilles, the appearance of the ceiling may be less cluttered; and (c) in conjunction with the right sort of heating and ventilating equipment, it may be possible to extract the heat from the luminaires and re-use it. These ideas are further discussed in Chapter 9.

**20.1.4** Office accommodation associated with factories is usually of limited extent, and generally high quality treatment of the decor and lighting may not be justified; but, even with a limited budget, these days there is little difficulty in providing lighting of good quality and suitable for the area using readily available standard equipment. For prestige offices, where the decor is important, standard false-ceiling systems with recessed luminaires may be used, together with decorative features to give the interior some distinction[55].

**20.1.5** Reference should be made to the *CIBS/IES Code*[5] for detailed recommendations for lighting various kinds of offices. The differences in illuminances proposed for the various office spaces are justified by such factors as: 500 lux for ordinary general offices, needed to enable work to be done all over the space, with the occupants facing any direction, close to or distant from a window; 750 lux for open-plan offices, considered necessary as an amenity in this kind of working environment, even if the work is not very demanding visually; 750 lux for conference rooms, this is justified by the need to provide a concentrating and alert atmosphere, but it may be desirable to be able to change the lighting level (and even the colour-appearance of the lighting) to suit the type of meeting and time of day. For example, lowering the lighting will make a better atmosphere for a semi-social occasion, while bright direct lighting may help to keep conferees awake in that difficult hour just after lunch. A very flexible method of lighting conference and committee rooms is to use a *lighting raft* (Figure 20.1). This is a simple structure which can be constructed in-house, and carries several systems of lamps; for example, downlighters to illuminate the conference table, indirect lighting for a soft lighting effect across the room, and some directional lighting for

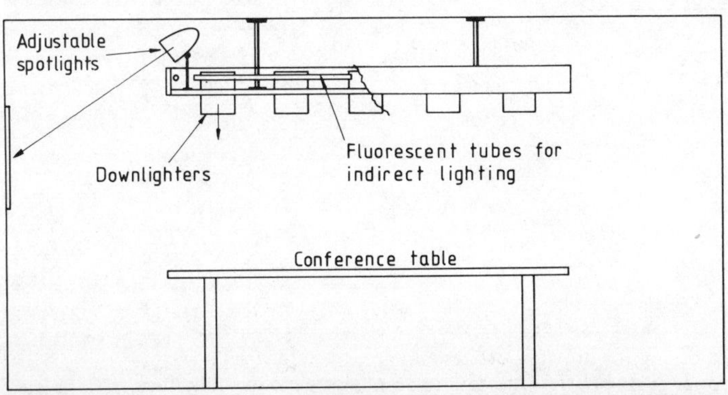


*Figure 20.1* Construction of a 'lighting raft' — a lighting fitment that is appropriate in conference rooms and boardrooms.

decor. All three circuits can be placed on separate dimmers to produce a highly flexible and pleasant means of lighting a room for meetings, committees, conferences, lectures etc, as well as for social occasions. If correctly dimensioned, it will present an appearance of 'floating' in the space, as the down-rods which hang it from the ceiling cannot be seen from normal angles of view.

## 20.2 Lighting for drawing-offices

**20.2.1** The *CIBS/IES Code*[5] recommendation of 750 lux general lighting for drawing-offices can be variously interpreted according to the type of drawing-office and the numbers employed there. For smaller offices, the room may be lighted to 500 lux of general lighting, and adjustable lights provided at each workstation to augment the general lighting under the control of the draftsman. This is a traditional method, and often preferred by the staff; in some cases these days, the adjustable task-light over the drawing-baord may be fitted with two or more miniature fluorescent tubes instead of the more commonly used filament lamp. Large drawing offices, say where there are 20 or more drawing-baords, and particularly those in which boards larger than A0 size are used, may benefit from the use of a special lighting system designed to give adequate illuminance on boards inclined at angles far from the horizontal (17.2). One very good method is to use a *coffered ceiling*; this can be a proprietary ceiling and lighting system, or may be constructed *in situ* (Figure 20.2). Because of the large exposed areas of the cells in such a ceiling, it will be necessary to construct them of fire-resistant materials to the satisfaction of the insurers and Fire Prevention Officer; e.g. they may be constructed of sheet metal, asbestos board, or fire-resistant building board. The system consists of panels forming a pattern of ceiling cells by attaching vertical lamina to the structural ceiling, so that the ceiling is covered with open-bottom cells, each containing one or more fluorescent batten luminaires. The cut-off angle is 70° from the vertical. The ceiling surface is

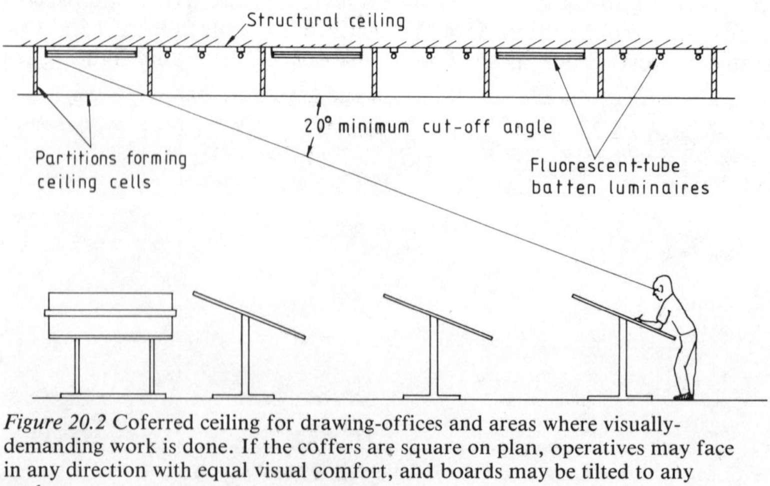


*Figure 20.2* Coferred ceiling for drawing-offices and areas where visually-demanding work is done. If the coffers are square on plan, operatives may face in any direction with equal visual comfort, and boards may be tilted to any angle.

painted matt white; the walls of the cells may also be white, or perhaps a pale pastel tint of 0.6 to 0.85 reflection factor. If fine colour judgment has to be exercised in the course of the draftsmen's work, then better colour-rendering tubes will have to be used (Chapter 4), and the walls of the ceiling cells will have to be coloured white or a pale shade of a grey devised from a mixture only of black and white. The cell system described will give a high cylindrical illuminance (17.2), but may be so diffused that it may be rather difficult to see the minute holes in drawing materials made by compasses and dividers; however, it should not usually be necessary to use local board-lights with this system, for the draftsmen and tracers soon learn to mark compass point centres with crossed construction lines. The size of the cells will depend on the length of fluorescent tubes to be used, and the height of the ceiling; but with the availability of efficient tubes of 1200 mm and 1050 mm at 40 W loading, cells of 1500 m and 1350 mm square respectively should be considered.

## 20.3 Lighting for canteens, staff rooms and clinics

**20.3.1** Luminaires for use in the kitchens and in the areas where food is prepared or stored should be of the types discussed for use in the food industry (19.1). Even in small establishments, the provision of pleasant lighting in the area where food is consumed will be much appreciated by the staff, and lighting of the whole area by lamps of good colour-appearance to an illuminance of 100 lux or more will tend to promote cleanliness; indeed, an illuminance of 300 lux would be even better. If there is a senior staff or directors' dining room, this might be given a rather more comfortable decor, tending towards a domestic rather than an institutional appearance. Note that as dining rooms are used for quite short hours, only a notional saving in energy will be achieved by lighting them with fluorescent tubes; indeed, the familiar colour of light from tungsten-filament lamps is preferred by most people for eating by, and the capital cost of the luminaires will be smaller too. A small expenditure on some lighting track upon which may be mounted a few adjustable spotlamps to emphasize a picture or some indoor plants etc, will greatly enhance the feeling of comfort and relaxation desirable in these areas where food and drink is consumed.

**20.3.2** Staff rooms vary very much according to their size and function, some serving also as a dining area or works club and bar, while others are used for purposes such as staff training lectures as well as for social functions. The choice of lighting system should take account of such variable uses, and the provision of some flexibility in the lighting control (including the facility to dim some or all of the luminaires) will increase the utility of the lighting.

The writer can reflect on the hundreds of factories he has visited during his career, recalling that an almost invariable sign of poor management has been dirty toilet areas, usually with poor lighting or lighting that was neglected or vandalized. There is cause and effect here; poor lighting will engender lack of respect for the surroundings, so that the cleaners cannot take pride in them, and the users will leave mess and dirty basins etc after using the facilities. Well lighted lavatory and washing areas are definitely conducive to better hygiene, and to better respect for the premises.

The cost of lighting these areas, including providing a good light at the mirrors, is so small that economy is pointless. If there is a problem with damage to the luminaires, or stealing of lamps etc, then fit vandal-resistant lighting equipment as is now used in public toilets; then, if the lighting equipment is intact, the provision of good illumination will tend to promote cleanliness and care of the facilities.

**20.3.3** Clinics and first-aid rooms need suitable lighting for hygienic purposes, and also to facilitate diagnosis and treatment. The use of better-colour-rendering lamps is necessary, as the complexions of persons is an important diagnostic aid. Enclosed, dust-tight luminaires would be appropriate (so that the whole room can be stringently cleaned when necessary), and a wheeled pedestal lamp as is used in doctors' surgeries will be a considerable asset when dealing with foreign bodies in eyes or extracting small splinters from the skin. If there is a rest-bay, the ability to dim down or switch down the lighting level so a sick person can sleep is desirable. In the treatment room a general illuminance of 300 lux is needed, with 500 lux where patients are examined; these levels can be readily obtained in small surgeries by localizing suitable luminaires to the examination area. Ophthalmic wall-charts and near-vision tests require 500 lux; dimmed lighting to 100 lux would be appropriate where vision-screening (3.1.5) is performed.

## 20.4 Stairs, corridors, circulation areas and entrances

**20.4.1** Where a staircase forms part of the reception area for visitors, it is tempting to provide lighting that is rather more decorative than functional. The safety aspects of lighting for stairways are important, for falls on stairs account for a significant number of industrial injuries. In modern buildings, where there are open ballustrades and staircases without risers to the treads, the choice and positioning of the luminaires should take account of possible glare being presented to the person on the staircase. Where the risers and treads are of uniform colour, it may be sound practice to try and leave the risers somewhat shadowed by the positioning of the luminaires on landings and half-landings, but bearing in mind that this can be glareful to a person mounting the stairway (Figure 20.3). If the emergency lighting is to consist of 'maintained' or 'sustained' stand-by luminaires (Chapter 6) the positioning must be arranged to ensure that safe movement will be facilitated when only the emergency units are in operation. The lighting for stairs should integrate with lighting in the adjacent corridors and circulation areas, but the facility of two-way switching is not recommended for stairs which may be traversed by visitors who could have difficulty in finding the switches. The use of pilot lighting (which cannot be switched off when the building is occupied) will make for safety on stairs, the pilot lights being also the emergency lights, e.g. 'maintained' or 'sustained' emergency luminaires (6.1).

**20.4.2** In the *Code*[5] the only areas for which scalar illuminance (17.2) is recommended are those for circulation of persons within a building. The objective is to provide the appearance of a 'well lighted space' (though, of course, it is only the planes that define a space that can be luminous) in which presons can move safely and in which they feel comfortable. The provision of a flow of light in virtually all directions means that the lighting will be substantially shadowless. Section A1.4 of the *Code* gives the calculation

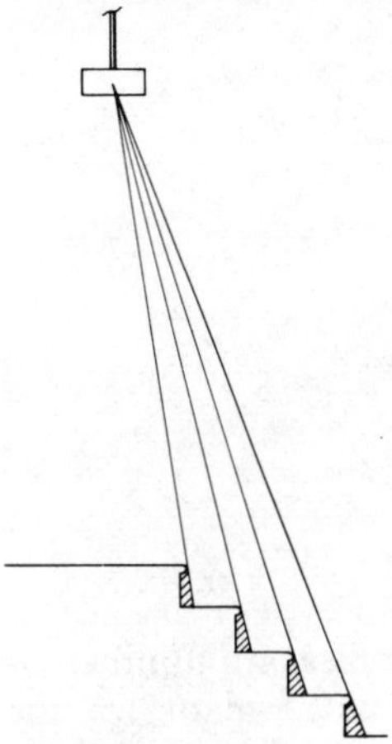

*Figure 20.3* Lighting stairs. Positioning a luminaire to give some shadowing of the risers can make for safety by enabling the toes of the treads to be seen in greater contrast.

method, and satisfactory results are readily achieved by use of general-diffusing luminaires (preferably suspended below ceiling level, and giving a substantial proportion of upward light) and the use of light-colured decoration for walls and ceiling. As a proportion of the cost of the whole lighting installation in a factory or office block, the lighting for the corridors, circulation areas and entrances can hardly be significant, and, as such a beneficial impression is gained by visitors when these areas are well-lighted, the exercise of some extra care in design and a modicum of extra expenditure would seem to be well justified. It is not necessary for the lighting to be 'decorative'; but rather that it shall be adequate and tasteful. A plain decor can be uplifted to give a most favourable impression by the use of some track with a few spotlights to illuminate a picture or two or some plants or flowers. Good lighting, with some carefully chosen areas of coloured surfaces, can be excellent decor in itself. Thus, the entrance area may be decorated in plain tasteful colours, and illuminated to a moderate illuminance, say 200 to 300 lux by enclosed prismatic-diffusing fluorescent luminaires. These may be mounted close-ceiling at heights up to around 3.5 m, depending on the size of the room; in smaller areas, it may be preferred to suspend the luminaires at around 2.5 m above the floor. A few track-mounted spotlights may be used to preferentially light the seating area, the reception desk and to highlight some colourful non-essential pictures, plants etc. If armchairs and a coffee table are provided in the waiting or interview area, one or two modern floor-standards or table lamps (of kinds used in one's home) will be appropriate to create an informal and relaxed atmosphere.

**20.4.3** The possible need for gradations of illuminance to deal with the transition between outdoor and indoor lighting conditions has been noted for industrial areas (Paragraph 2.3, Figure 2.2), and the same principles may be applied to office entrances and the prestige entrance for visitors. The visitor arriving by foot or by car should be steered by the pattern of lighting towards the visitors' entrance, and, without extravagant cost, he may be presented with a welcoming and pleasing approach to the organization. First impressions of organizations, like those of people, are often lasting and tend to shape the subsequent relationship. The small cost of a well-lighted entrance and foyer, with good signing and reception, may be well repaid by the benefit to the prestige of the organization.

Appendix I

# UK legislation on industrial lighting

**App. I—1** In the United Kingdom, and in an ever-increasing number of other countries, the law requires that certain provisions be made for the health and safety of occupants of business and public premises; and these provisions include 'sufficient and suitable' lighting. This Appendix reviews the duties of occupiers of premises under British legislation and Common Law to ensure the reasonable safety of persons. It is believed by the author that the guidance given throughout this book conforms to the letter and the spirit of UK legal requirements; further, it is believed that compliance with these recommended procedures will also satisfy the requirements in many other countries. The user should satisfy himself that by following these recommendations he will not bring himself into a breach of local laws by any act of commission or omission. Because of harmonization of laws that is taking place within the member countries of the European Economic Community (EEC), and the technical harmonization now occurring world-wide by the work of the *Commission Internationale d'Eclairage* (Appendix II), differences of practice between countries, where they exist, are not usually significant, and are likely to become less with the passage of time.

**App. I—2** Apart from legislation which imposes specific duties upon occupiers of premises, the British Common Law embodies a general principle that the occupier must take steps to ensure the reasonable safety of persons who enter his premises, this being capable of wide interpretation in the courts. The former *Factories (Standards of Lighting) Regulations (1941)*, which imposed certain very low minimum requirements for illuminance in specified industrial premises, have been replaced by the more general requirements of the *Health & Safety at Work Etc Act* (1974) which imposes a general duty on employers and others to provide for the reasonable safety of others; this can be construed to include needing to provide 'sufficient and suitable' lighting to ensure the reasonable safety of employees and others entering the premises. The Act does not cite specific illuminances, but enables recognised codes of practice to be adopted (in this and in other matters). The recognised code of practice for lighting is the *CIBS/IES Code for Interior Lighting*[5], but it should always be borne in mind that the Health & Safety Executive officers have discretionary powers, and may accept as sufficient and suitable an illuminance that was lower than that recommended in the *Code*. Of course, the *Code* recommendations are not enforceable as such; it sets standards of good practice as recognised by professional advisors and which are derived from long experience and much research. If there was a legal dispute as to whether lighting in a place was sufficient and suitable, it would be good proof that it was so if it could be shown it complied with the recommendations of the *Code*. Similarly, under Common Law, if an injured person (or the estate of a deceased person) sought damages in relation to an accident resulting in injury or death said to be due to insufficient or unsuitable lighting, proof that the lighting in fact complied with the *Code* would be a good defence against an accusation of negligence.

**App. I—3** It is a requirement of UK law that persons shall be able to escape from premises when danger threatens—even if the normal lighting has failed due to any cause. The *Fire Precautions Act* (1971) requires that the exit routes shall be capable of use at all times; and this is construed to mean that sufficient and suitable lighting must be provided along those routes to enable them to be used, and to be used in reasonable safety. The provision of the normal lighting does not discharge this duty, for the blame for failure to provide the lighting cannot be referred to another party; thus the occupier may need to install emergency lighting (Chapter 6). Between 1971 and 1974, the standards of emergency lighting as applied by the Enforcing Officers under the Fire Precautions Act were not uniform throughout the country; but the issue of *BS 5266: Part 1: Emergency Lighting*[24] provided a code of practice which could be used to interpret the requirements of the Act. Certain criticisms of the BS have been made in Chapter 6, but these do not impair the importance of the BS in this role. The Enforcing Officers under the Fire Precautions Act do have discretionary powers; for example, sometimes the Officer will rule that emergency lighting in particular premises need not be provided, on the grounds that sufficient ambient light will result from adjacent public lighting. This could be a ruling leading to danger, if the failure of the lighting was due to failure of the public supply, with the result that the adjacent public lighting was also extinguished. The occupier would be well advised to protect the occupants from personal danger, and himself from risk of prosecution, by following the guide-lines given in Chapter 6.

**App. I—4** It should be noted that these requirements for lighting to provide for the safety of persons apply not only to completed premises, but also to those under construction or in the course of alteration. Thus, persons actually installing lighting and emergency lighting must themselves be provided with sufficient and suitable lighting, and shall be able to use the exit routes at all times. Further, the term 'premises' includes exterior spaces as well as the interiors of buildings, and some emergency exit routes may themselves not be within buildings. The objective of an exit route will be a 'designated place of safety'; if the route out of a burning building was lighted, but the lighted route led only to an adjacent exterior which itself had a high fire risk, then the letter and spirit of the law would not have been complied with. Thus, emergency lighting may have to be provided along exterior escape routes as well as those within buildings[1].

**App. I—5** Another important law is the *Offices, Shops & Railway Premises Act (1963)* which, in a similar way to *Health & Safety at Work Etc Act,* makes a requirement that the occupier provides 'sufficient and suitable lighting' (App. I—2) necessary for the welfare and safety of his employees. Again, the *CIBS/IES Code*[5] serves the purpose of a code of practice which is recognized in the courts as defining what is sufficient and suitable in this context.

**App. I—6** The term 'suitable lighting' may be taken to mean lighting that was not so lacking in uniformity as to be a potential cause of danger; and that it would not cause disabling glare which would handicap the vision of anyone working in the area. Further, the definition could include suitability of colour—particularly of colour-rendering; this would be of considerable importance in an area where colour-coded wires, metal stocks or containers of gases were used, or where colour-coded instructions of any kind must be followed for safety purposes (4.1). In such a workplace, the use of *low-pressure* sodium lighting would not be suitable, for it has no colour-rendering property. Because of this lack of colour-rendering, the author considers these lamps to be unsuitable for illuminating any workplace, and they have therefore not been included in the summary of lamp data (Appendix III).

**App. I—7** Where light spreads outside the boundaries of the premises it may cause nuisance or danger. Nuisance from light may include that due to light entering the bedrooms of adjacent homes, light which affects livestock (e.g. poultry), or that which affects plants (e.g., causing early blooming of greenhouse plants) etc. In all of these, the person suffering the nuisance or damage may claim for compensation in the courts, and the courts may place constraints upon the offender to cease or ease the nuisance. There are statutory restraints on causing danger by extraneous light which may affect vehicular traffic outside the premises, or trains (glare to drivers, confusion with signals), and specifically against showing a light to seaward causing danger to mariners or 'confusible with a navigation light', this constraint also applying to some rivers and inland waterways.

**App. I—8** In general, the illuminances required to ensure the safety of persons when moving about premises are far lower than the recommended illuminances of the *CIBS/IES Code.* The illuminances required in occupied areas to promote the health and welfare of occupants may be greater than those required for movement, but again are less than the Code's recommendations. Thus, the occupier will be prudent if he lights his premises in accordance with the *Code,* when he may be confident that he will also have satisfied the Law's demands as regards the provision of sufficient and suitable lighting (Appendix II).

Appendix II

# Summary of CIBS/IES Code recommendations

**App. II—1** Throughout this book, and particularly in this Appendix, frequent reference is made to the *CIBS/IES Code for interior lighting*[5]. For nearly fifty years, the former Illuminating Engineering Society, and its successor the Chartered Institution of Building Services (Lighting Division) have brought out successive editions of this invaluable guide to good lighting practice. Through the years, the recommendations have evolved through many changes; originally, illuminances were specified in *foot-candles* (1 fc = 1 $lm/ft^2$ = 10.76 lux). Recent editions have included recommendations as to *limiting glare index* and choice of light-sources for suitable colour properties as well as for illuminance. The editions have reflected the changes in lighting practice due to technical development of sources, and changes in architectural practice and the lighting needs in industrial processes and office applications. The 1968 edition was the last to specify illuminance recommendations as 'minimum service illuminance'; the 1973 and 1977 editions quote instead 'standard service illuminance', a basic recommendation that can be adjusted upward or downward according to particular circumstances. While some information on the *Code's* recommendations is given here, no summary or condensation can equal the full text which contains a great deal of wise guidance; therefore the reader is recommended to obtain a copy of the current *Code*—and to take note of the whole text, not just the tables of illuminance recommendations.

**App. II—2** There have been times in the past when the *Code* has been attacked by critics who have accused the compilers of advocating wastefully high illuminances; and it has even been suggested that the *Code* was a means of promoting sales of lamps and lighting equipment. An argument put forward was that, if the basis of the *Code* is sound, it would not have been necessary to increase the illuminance recommendations through successive editions. Having first stated that the former IES and the CIBS have always been strong advocates for the wise and economical use of energy, and refuting utterly any suggestion that lighting manufacturers have ever influenced the *Codes*' compilers to promote excessive lighting levels, some explanation of why the recommendations have changed over the years may be given. Each edition of the *Code* has embodied the results of careful study by responsible officers of the learned society concerned, taking into account the economics of light-sources available at each time, giving due weight to the known scientific facts relating to vision and the efficient performance of work, and—increasingly—taking account of the practices current and the preferences of users. These matters could be explained in detail and at great length; but it should suffice for the general reader to have these assurances, so he may be confident that in following the *Code's* recommendations he will be applying the results of the thorough studies which will lead to acceptable standards of visual performance, to safety, and to good economics of finance and energy.

**App. II—3** The relationship between the task illuminance and the visual performance of the workers is clearly established (Chapter 1). A matter which gives rise to some confusion is the effect of providing higher or lower illuminances than the *Code's* recommendations. Would it, for example, seriously handicap the worker if the illuminance was, say, 10 per cent lower than the recommended figure? The answer is: probably not to any noticeable extent, at least in the short term; but, for older workers (whose need for illuminance is somewhat greater), and for those parts of any task which approach closely to the limits of visual resolution of the worker, the lower illuminance will be of some handicap. Conversely, illuminances higher than the recommendations will not necessarily bring about any enhancement of task performance, particularly in the short term; but, well engineered lighting at levels somewhat higher than the standard service illuminances can only be beneficial. A sense of proportion is necessary here; one must remember that these days the user gets something rather more than four times as much light per unit of cost (viz, lumen-hours per £ or per dollar) than 30 years ago. The illuminances currently employed would have been totally impracticable in former times when only low-efficacy sources were available. It must also be noted that the total cost of lighting compared with other overheads, or as a percentage of the total cost of manufactured goods is just a fraction of a penny per £ of goods sold (Chapter 18). Because of the continued improvement in the efficacies of light-sources, the currently recommended illuminances may often be adopted with considerable economy in energy. Heat dissipated by luminaires can make a valuable contribution to building heating (Chapter 9), but this would not be a justification for using less efficacious sources or illuminances far above the *Code* recommendations.

**App. II—4** During the preparation of this book (1980/81), a further edition of the *Code* is in preparation, with a projected publication date in 1982. Recommendations embodied in Table 13 are therefore based on the currently published data, and may be subject to minor amendment on publication of future editions of the *Code*. The illuminance levels quoted in this Appendix are based on the 1977 Edition of the *Code*, plus the publications of The Electricity Council. The steps of illuminance used by the Code are generally those advocated by the Commission Internationale d'Éclairage, but it may be noted that in any representation of the illuminance steps on a linear scale against equal comparative steps, the step of 100 lux seems to fit the curve better than the conventional 150 lux (Figure II.1). (Illuminances of 50 lux and 25 lux are widely used for lighting farm buildings.)

**App. II—5** The *standard service illuminance* (Table 13) for each application or location must be adjusted for special conditions, e.g. if the reflectances or contrasts in the visual task or unusually low, if errors will have serious consequences (i.e. danger to personnel, great cost), if the task is of long or short duration, and if the area is windowless. It is the author's further

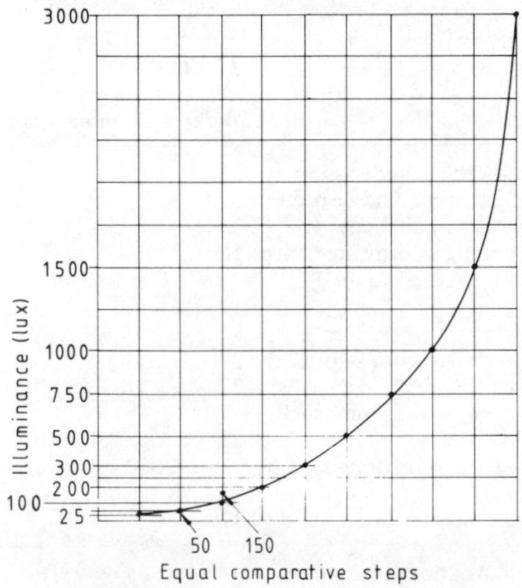


*Figure II.1* Steps of illuminance used in lighting recommendations examined against equal comparative steps.

recommendation that the illuminance should be further adjusted if eye protection must be worn continuously. The adjusted illuminance is termed the *final design illuminance* (Figure II.2).

**Table 13** Examples of illuminance recommendations, based on CIBS/IES Code[5]

| *Class of visual task* | *Typical examples* | *Limiting Glare Index* | *Standard Service Illuminance* |
|---|---|---|---|
| STORAGE AREAS AND PLANT ROOMS WITH NO CONTINUOUS WORK; movement and casual seeing | Boiler houses, pump houses | 25 | |
| | Stairs, gangways (steelworks) | — | |
| | Loading bays (large materials) | 25 | 150 |
| | Food stores (kitchens) | — | |
| | Rest rooms | 19 | |
| CASUAL WORK WITHOUT DIFFICULT VISUAL TASKS; movement and casual seeing | Assembly shops, casual work | 25 | |
| | Generator station turbine houses | 25 | |
| | Automatic processes (general lighting) | 25 | 200 |
| | Warehouses — small materials racks | 25 | |
| | Chemical raw materials stores | 25 | |
| ROUGH WORK: rough machining and assembly; simple tasks with large detail | Glassworks — mixing rooms | 25 | |
| | Steelworks — mould preparation | 28 | |
| | Laundries — receiving, sorting, washing | 25 | 300 |
| | Leather works — general lighting | 25 | |
| | Office print rooms | 19 | |

**Table II.1** — *continued*

| *Class of visual task* | *Typical examples* | *Limiting Glare Index* | *Standard Service Illuminance* |
|---|---|---|---|
| ROUTINE WORK; ordinary work with average detail | Engine and vehicle body assembly | 22 | |
| | Aircraft fabrication and inspection | 22 | |
| | Working areas — kitchens | 22 | 500 |
| | General lighting — drawing offices | 16 | |
| | General offices, clerical, typing | 19 | |
| DEMANDING WORK: fairly difficult visual tasks with small detail | Printing machine room — presses | 22 | |
| | Stores issue counters | 22 | |
| | Woodworking — medium bench and machine work | 22 | 750 |
| | Drawing office — on drawing boards | 16 | |
| | Deep-plan general offices | 19 | |
| FINE WORK; colour discrimination; very difficult visual tasks with very small detail | Telecommunications equipment inspection | 19 | |
| | Paintworks — colour matching | 19 | |
| | Fine bench and machine work | 22 | 1000 |
| | Printed sheet inspection | 19 | |
| | Jewellery — fine processes | 16 | |
| VERY FINE WORK; extremely difficult visual tasks with minute detail | Fine, intricate gauging and inspection | 16 | |
| | Gem cutting, polishing, setting | 19 | |
| | Upholstery (cloth inspection) | 16 | 1500 |
| | Hand-tailoring. Fine die-sinking | 19 | |
| | Fine soldering, spotwelding (instrument) | 19 | |
| MINUTE WORK: exceptionally difficult visual tasks with minute detail | Inspection (small instruments) | 19 | |
| | Jewellery, watchmaking (finest work) | 19 | |
| | Final inspection, perching (textiles) | 19 | 3000 |
| | Critical colour-matching | 19 | |
| | Close work with loupes and magnifiers | — | |

Reference to the *Code* itself is recommended to obtain detailed recommendations for most industrial locations and processes, together with guidance as to the position of measurement, the colour-appearance and type of lamp required, and additional notes. The *standard design illuminances* given in the *Code,* of which a few examples are quoted in this Table, must be adjusted to arrive at the *final service illuminance* (Figure II.2).

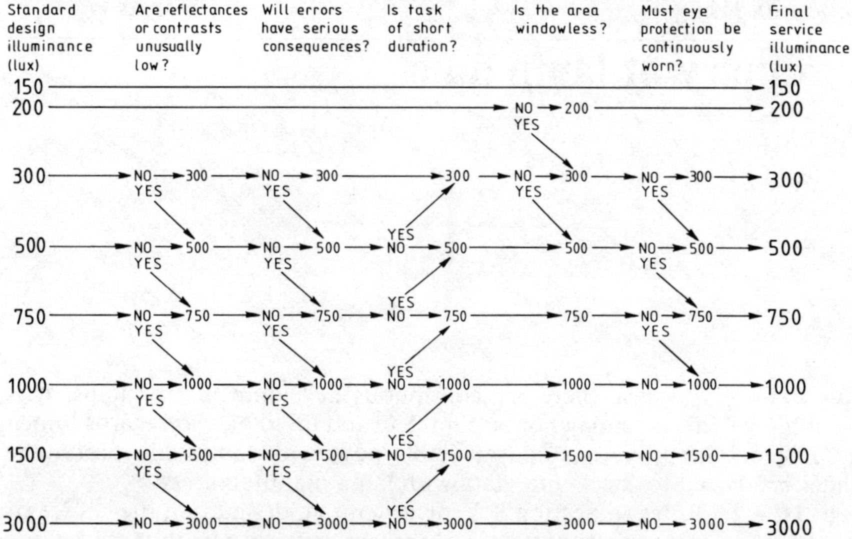


*Figure II.2* Flow-chart showing how the Standard Design Illuminance is to be adjusted to arrive at the Final Service Illuminance. This Table is based on a table in the 1977 *CIBS/IES Code*(5) but the adjustment for wearing eye protection continuously is the author's recommendation. The higher illuminances (1000 lux and above) will in most cases be obtained on the task by local lighting. Where necessary, the lighting will be supplemented by optical aids, loupes, magnifiers etc (Chapter 3).

Appendix III

# Summary of lamp data

**App. III—1** Because there are continuous developments in lamps, it is possible that this data may not be complete and up to date as regards lumen outputs, colour properties and available powers of lamps; therefore, the reader is advised to check vital data with lamp manufacturers.
**App. III—2** Refer to Section 8.1 for general guide-lines on the choice of lamps, and to the numerous sections of the text that deal with the applications of lamps for specific lighting tasks, noting in particular the factor of colour performance (4.2) and energy use (9.2). For general descriptions of the functions of the various types of lamps and their technical details, refer to another book by the author(1).
**App. III—3** As well as development in lamps, there is currently much progress in improving control gear. For example, with modern ignitors, SON lamps can now be provided with rapid restrike time after a brief power interruption. But also note the difficulties which can arise from attempting to use HID lamps on control gear for which they are not fully compatible (13.4).
**App. III—4** While all the types of lamps listed in this Appendix have some application to industrial lighting, some are better than others for specific uses. Fluorescent tubular lamps and HID lamps (SON, MBF, MBI) have higher efficacy (lumens per watt) than filament lamps and tungsten halogen lamps, and give far longer lives; but their capital cost is higher. According to the control gear used, HID lamps may take up to 3 to 8 minutes to relight after being switched off (but see App. III—3).
**App. III—4** In the following tables, only those types of lamps commonly used in industrial lighting are included. The types of lamps are identified by letter symbols used by lighting manufacturers in the UK, and these may differ in other countries, and sometimes even from one maker to another. The lumen outputs given are 'lighting design lumens' (16.2.7) at 220/240 V unless otherwise stated. The colour-appearance (4.2.2) is stated as 'warm', 'intermediate' or 'cool'. Colour rendering is stated as 'very good' (CIE Group 1, with Ra of 85 to 100), 'good' (CIE Group 2, with Ra of 70 to 84) or 'moderate' (CIE Group 3, with Ra of 50 to 69). For fluorescent tubes, only data for WHITE and triphosphor tubes are given in Table 16, but the properties of a number of other colours of fluorescent tubes are given in Table 5 (4.3.5). In describing the appearance of colours under a lamp, the colours are abbreviated thus: R: red; O: orange; Y: yellow; G: green, B: blue. The lamp shapes are coded thus: 'E': eliptical bulb; 'T': tubular bulb; 'L': linear bulb; 'R': reflector bulb.

**Table 14** *Type SON series lamps.* Son Improved colour high-pressure sodium-vapour high-intensity discharge (HID) lamps

| Watts | 50 | 70 | 150 | 210 | 250 | 310 | 360 | 400 | 600 | 1000 |
|---|---|---|---|---|---|---|---|---|---|---|
| Lumens (E) | 3300* | 5800* | 1400 | 17250 | 25000 | 33000 | 36000 | 47000 | | 110000 |
| Lumens (T) | | | 14500 | | 25000 | 34500 | 38000 | 48000 | 65000 | 123000 |
| Lumens (R) | | | | | 18900 | 24000 | 27000 | 33000 | | |
| Lumens (L) | | | | | 24000 | 34500 | | 46000 | | 125000 |

* Initial lumens.

*Nominal life:* 8000 h, but guarantees of up to 20,000 h are given by some manufacturers. Very long life is obtained in installations with infrequent switching.
*Efficacies:* typically 63–110 lm/W including control-gear losses.
*Colour-appearance:* warm; golden colour-appearance, but lamps with whiter appearance becoming available.
*Colour-rendering:* Bs appear violet; Gs slightly yellow; Ys very bright, slightly O; Os and Rs bright, Rs slightly Y. Colour-rendering index, around 25 (Moderate).
*Approx. correlated colour-temperature:* 2100 K.
Note: An improved type of lamp with slightly lower efficacy and much improved CRI (around 85) is currently becoming available, initially in 250 W size.

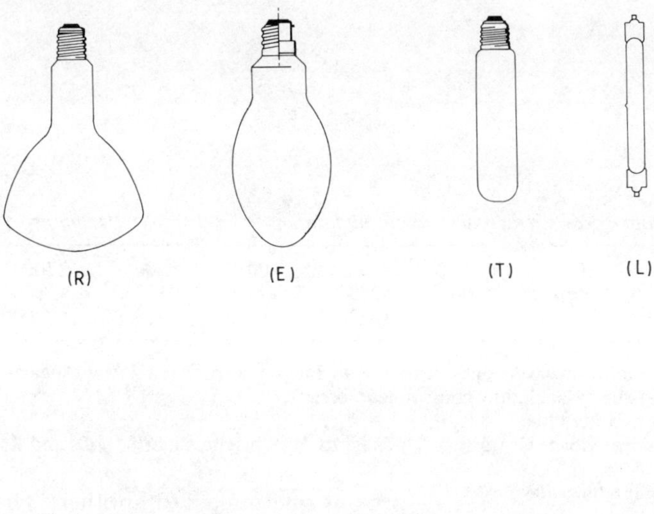

**Table 15** *Type MBI series lamps.* MBI metal halide lamps; improved colour-rendering mercury-vapour high-intensity discharge (HID) lamps. Also designated as 'mercury halide'. Symbols HPI and HQI are also used for this type of lamp.

| | | | | | |
|---|---|---|---|---|---|
| Watts | 250 | 400 | 750 | 1000 | 2000 |
| Lumens (E) | 16000 | 24000 | | 85000 | |
| Lumens (T) | | | | | 150000 |
| Lumens (L) | | | 60000 | | |

*Nominal life:* 6000 or 7500 h; but somewhat longer life may be obtained in installations with infrequent switching.
*Efficacies:* typically 63–72 1m/W including control-gear loss; efficacy is affected by the burning position, eg. normal attitude is cap-up. Lamps designed for cap-down or horizontal operation may give lower outputs.
*Colour-appearance:* Cool, blue-white appearance.
*Colour-rendering:* Gs appear bright, slightly Y; Ys very bright, slightly O; Os and Rs, medium. (Superior).
*Approx. correlated colour-temperature:* MBIF 3900 K, MIBI/MBIL 3600/4400 K. A fluorescent-coated version (MBIF) is available in some powers.

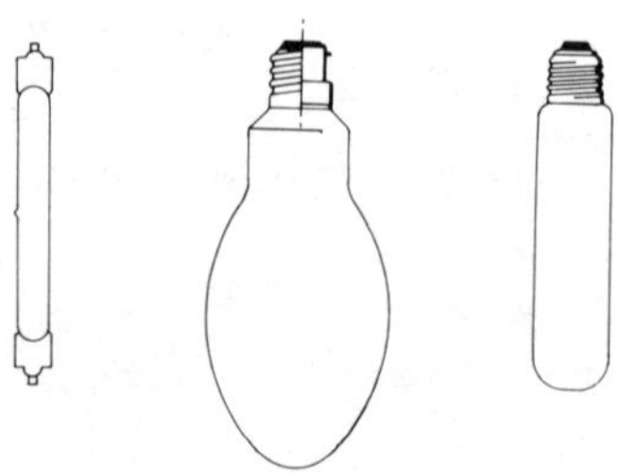

(L) (E) (T)

**Table 16** *Type MBF colour-corrected mercury-vapour high-intensity discharge (HID) lamps*

| | | | | | | | |
|---|---|---|---|---|---|---|---|
| Watts | 50 | 80 | 125 | 250 | 400 | 700 | 1000 |
| Lumens (E) | 1900 | 3650 | 5800 | 12500 | 21500 | 38000 | 58000 |
| Lumens (R) | | | | 10500 | 18000 | 32500 | 48000 |

*Nominal life:* 7500 h, though lamps will continue to operate long after this but at lower efficacy.
*Efficacies:* typically 35–50 lm/W including control-gear losses.
*Colour-appearance:* cool: rather blue.
*Colour-rendering:* Bs appear violet; Gs fairly bright, slightly Y; Y bright, slightly G; Os and Rs fairly dull. (Good).
*Approx. correlated colour temperature:* 4000 K.

(R) (E)

**Table 17** *Type MCF fluorescent tubular lamps (white and triphosphor colours)*

| *Watts* | 4 | 6 | 8 | 13 | 15 | 15 | 18 | 20 | 30 | 36 | 40 |
|---|---|---|---|---|---|---|---|---|---|---|---|
| *Length* | | | | | | | | | | | |
| 38 mm dia: | | | | | | 450 | 600 | 600 | | 1200 | 600 |
| 26 mm dia: | | | | | 450 | | | | 900 | | |
| 16 mm dia: | 150 | 225 | 300 | 525 | | | | | | | |
| *Lumens* | | | | | | | | | | | |
| white | 100 | 250 | 420 | 750 | 800 | 750 | | 1100 | 1850 | | 1700 |
| triphosphor | | | | | | | 1325 | 1200 | 2250 | 3200 | |

| *Watts* | 40 | 40 | 50 | 58 | 65 | 75 | 80 | 85 | 85 | 100 | 125 |
|---|---|---|---|---|---|---|---|---|---|---|---|
| *Length* | | | | | | | | | | | |
| 38 mm dia: | | 1050 | | | 1500 | 1800 | 1500 | 1800 | 2400 | 2400 | 2400 |
| 26 mm dia: | | | | | | | | | | | |
| 16 mm dia: | 1050 | | 1500 | 1500 | | | | | | | |
| *Lumens* | | | | | | | | | | | |
| white | 2800 | 2800 | 3600 | | 4750 | 5750 | 5200 | 6250 | 6850 | 8000 | 8900 |
| triphosphor | | 3000 | | 4900 | 4900 | | | 6300 | 7200 | | |

*Nominal life:* Tubes of 20 W rating and above are usually rated at 7500 h, smaller powers at 5000 h. However, tubes will continue to operate long after this but at lower efficacy. Infrequent switching prolongs life.
*Efficacies:* for WHITE tubes 40 W and above, typically 62-66 lm/W, and for triphosphor tubes 69-70 lm/W, both including control-gear losses.
*Colour-appearance:* WHITE tubes, intermediate. Triphosphor tubes, cool, intermediate or warm according to type.
*Colour-rendering:* WHITE — Ra 56 (moderate); Bs dull, appear violet; Gs bright, slightly Y; Ys very bright, slightly O; Os and Rs bright, Rs slightly Y. Triphosphor — Ra 85 (very good); Blue/violet, green and orange/red are intensified, other colours slightly dull.
*Approx. correlated colour temperatures:* WHITE — 3400 K. Triphosphor — 3000, 3400 or 4100 K according to type. This table should be read in conjunction with Table 5 (4.3.5).
Reflectorised tubes are used for certain local lighting applications, and their technical data differs from that given for standard tubes. Reflectorised tubes are not recommended for general lighting purposes.

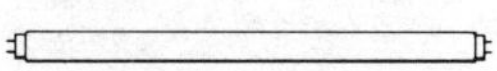

**Table 18** *Type TH tungsten-halogen lamps*

| | | | | | | | |
|---|---|---|---|---|---|---|---|
| Watts | 200 | 300 | 500 | 750 | 1000 | 1500 | 2000 |
| Lumens | 2800 | 5000 | 9500 | 14500 | 21000 | 32000 | 44000 |

*Nominal life:* 2000 h.
*Efficacy:* 14-22 lm/W. No control gear.
*Colour-appearance:* Warm, yellow/white.
*Colour-rendering:* Gs bright, slightly Y; Ys very bright; Os and Rs medium; Bs weak. (Good).
*Approx. correlated colour temperature:* 2800-3100 K. The 200, 300 and 500 W powers are of same physical dimensions.

**Table 19** *Type GLS tungsten-filament lamps*

Smaller powers:

| | (Coiled coil) | | | (Single coil) |
|---|---|---|---|---|
| Watts | 40 | 60 | 100 | 150 |
| Lumens | 390 | 665 | 1260 | 1960 |

40/100 W

Larger powers: (Single coil)

| | | | | | |
|---|---|---|---|---|---|
| Watts | 200 | 300 | 500 | 750 | 1000 |
| Lumens | 2720 | 4300 | 7700 | 12400 | 17300 |

150/200W

*Nominal life:* 1000 h.
*Efficacies:* Smaller powers 10-13 lm/W; larger powers 13/18 lm/W. No control gear.
*Colour-appearance:* Smaller powers, warmer and more orange/yellow than TH; larger powers, slightly cooler.
*Colour-rendering:* Larger powers, generally as TH, but more biased to O and R, with weak Bs; smaller powers, similar, but the Bs are weaker. (Moderate).
*Approx. correlated colour-temperatures:* Larger powers, 2700-2800 K; smaller powers, 2600-2700 K.
The use of these lamps for long hours of burning, for general lighting or for security lighting is not recommended.

300/1000 W

**Table 20** *PAR-38 pressed glass reflector floodlamp (reflectorised tungsten-filament lamp)*

| | | |
|---|---|---|
| Watts | 100 | 150 |
| Lumens | Not applicable | |

A 150 W PAR lamp directed at a surface 3 m away and normal to the beam axis will produce a circular area illuminated to at least 150 lux over a diameter of 3 m, with more than 300 lux at the centre. (See 17.4.2 for a method of measuring the illuminance in local lighting applications.)

Appendix IV

# Summary of luminaire data

## App. IV—1 General

**App. IV—1.1** Because of continuous developments in luminaire design, it is possible that this data may not be complete and up to date as regards performance and features of luminaires; further, because luminaires serving the same duty but made by different manufacturers may differ in constructional detail, the illustrations in this Appendix must be taken only as indicative of types of luminaires and not of specific products.

**App. IV—1.2** Throughout the book there are many guidelines to the selection of luminaires to suit particular lighting applications. Special attention should be given to selection of luminaires and all electrical enclosures to suit environmental hazards (Chapters 10, 11 and 12).

**App. IV—1.3** Because there are hundreds of individual designs of luminaires suitable for applications in industrial lighting, no attempt is made here to review all the types and makes of luminaires which might be employed. Instead, the main varieties of luminaires employed in industrial lighting applications are listed, giving an illustration of a typical luminaire in each category, with some notes on selection and use. Only luminaire types commonly used in industrial lighting are included in this Appendix; patterns more commonly used in commercial buildings have been excluded. Similarly, luminaires used primarily for exterior lighting have been summarized by the author in another book(1) and are not listed here.

## App. IV—2 Fluorescent lamp luminaires

**App. IV—2.1** *Bare batten luminaires.* Available to take one or two fluorescent tubular lamps of 20 W and above as listed in Appendix III (Table 17). Miniature batten luminaires are also available for some of the smaller

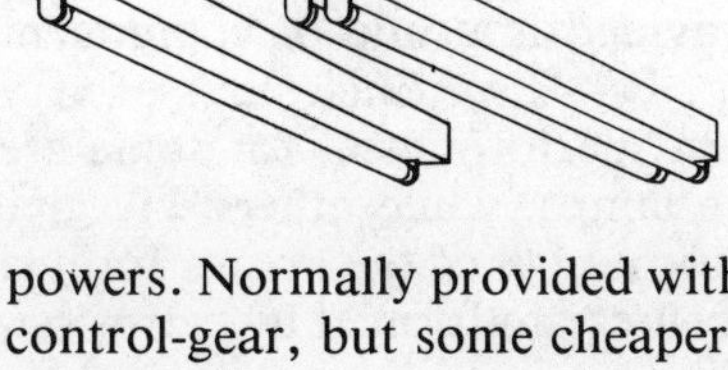

powers. Normally provided with power-factor-correction capacitor with the control-gear, but some cheaper patterns do not include capacitor. Control gear may be (a) switch-start (requiring a plug-in starter cannister switch) or (b) quick-start (start without a starter switch by virtue of a voltage kick produced by the control-gear circuit). Most batten luminaires form part of a luminaire range, e.g. reflectors or diffusers may be available to clip on to form other luminaire types. Twin-lamp fluorescent tube luminaires are designed for operation with both lamps, and some can overheat or develop faults if operated with one tube only. 'Lead-lag' twin-lamp luminaires are used to reduce stroboscopic effect (5.3.4). In small rooms having high reflectance ceiling and walls, bare-lamp luminaires may be acceptable, but they are not suitable for general lighting of large areas because they are glareful and not efficient. Although the tube delivers a very high proportion of its lumens into the space, a large percentage is received by the ceiling and upper walls, so the utilization factor (16.2) is poor; open-trough reflectors added to the battens will typically enable the lamps to deliver 25 per cent more illuminance on the working plane.

**App. IV—2.2** *Trough luminaires.* Available to take one or two fluorescent tubular lamps from 900 mm length and over as shown in App. III (Table 16). This type of luminaire has somewhat fallen from favour since the availability of cheap extruded wrap-around diffusers (see App. IV—2.3), but still has

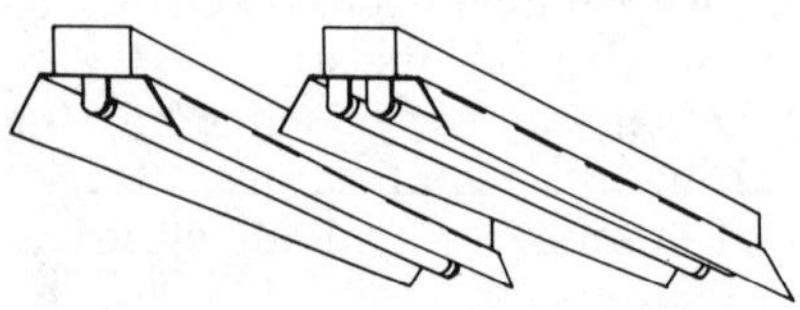

much to recommend it, being robust, easy to maintain, and providing positive cut-off of the view of the lamp (usually at 70° from the downward vertical). Trough reflectors are commonly made of sheet metal stoved white, but some are made of translucent white plastic sheet. Metal reflectors are better if slotted to permit some upward light and thus preventing tunnel effect (11.1.7) as well as having a self-cleaning effect (11.1.6). The best design is that in which the slots illuminate the batten and the upper sides of the reflector surfaces, thus reducing contrast. Some trough luminaires are made so that the reflector (which may be in one piece or as two 'wings') can be removed for ease of cleaning. Although the paint surfaces of the reflectors will depreciate permanently in time, these luminaires are simple to maintain, and cleaning can be performed rapidly. As can some other fluorescent-lamp luminaires, trough units can be mounted end-to-end to form continuous rows. In some patterns, means of end alignment is provided, with facility for through-wiring; in others, the unit is designed to attach to a trunking section. In the latter case, gear trays only (without enclosure) may be inserted into the underside of the trunking, thus saving weight and cost.

**App. IV—2.3** *Refractor and diffuser enclosed luminaires.* These refractor and diffuser units may be provided as complete purpose-made luminaires, or the refractor or diffuser enclosure may be provided as an add-on component to be fitted to a bare batten luminaire (App. IV—2.1). Enclosures of very similar appearance may give greatly different performance, for some are designed only to diffuse the light from the lamps, while others will give precise optical control because of the scientific design of the prisms formed on the wall of the enclosure (a). Some enclosures are formed by extrusion,

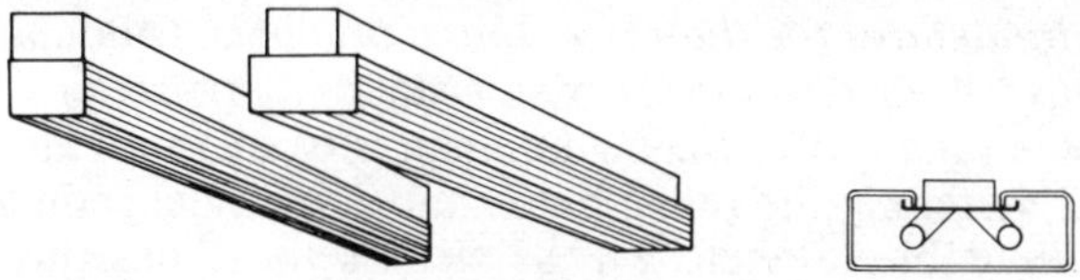

(a) (b)

and have longitudinal ribs only; others are formed by pressing, blowing or fabrication, and may have intricate systems of prisms formed for light control. Maintenance and cleaning is facilitated if the ribbing or prisms are formed on the inside of the enclosure. Some systems have larger enclosures for the twin-lamp batten, this giving lower brightness and better light control. Patterns vary very much, ranging from simple additions to battens (as illustrated) to designs in which the wrap-around enclosure conceals the batten spine (b). Very attractive units of this kind present shallow depth, and may accommodate as many as four fluorescent tubes. However, the utilization factor of smaller section enclosures is lower, and there will be losses of a few per cent of the output due to mutual light obstruction by the lamps. While most luminaires of this type are suitable for both conduit-suspension or close-ceiling mounting, some are only suitable for one or the other. Note that diffuser enclosures merely reduce brightness, while refractor enclosures may do this while directing light flux towards the desired directions for good utilization.

**App. IV—2.4** *Low-brightness and louvred luminaires.* Small-cell louvre grids, of white colour or of polished metal or diffusing plastic may be used to conceal the bright tubes, both giving 'cut-off' and controlling the light according to the optical design. In the simplest application of louvres for this purpose, an 'egg-crate' louvre is fixed in the base opening of a trough luminaire similar to those discussed in Appendix IV—2.2.

A variation of the general idea is employed in luminaires designed to give a high proportion of upward light but with reduced brightness at normal angles of view, as is achieved in a type of luminaire originally devised for use in hospital wards, but equally useful for lighting large office areas, as well as in canteens and restrooms (a).

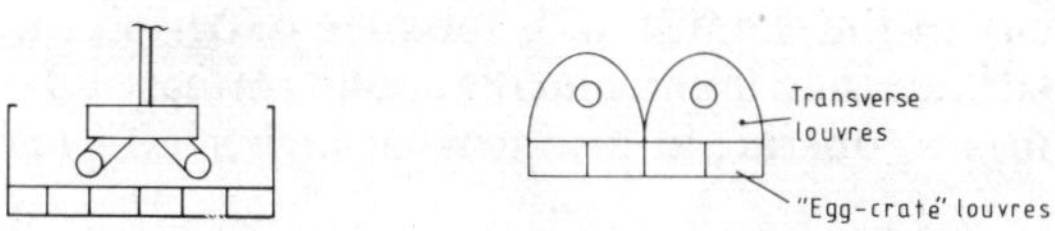


(a) (b)

If the louvres are made of specular material (plated and polished metal, or aluminised plastic) then very low brightness luminaires can result. A design of this kind is shown in principle in sketch (b), and can be utilized in the form of a suspended luminaire, a close-ceiling unit, or recessed into the ceiling. Although there is almost no glare at all from the luminaires, in installations using such equipment it may be found that the appearance of the room is gloomy and unsatisfying, even though the illuminance may be of several thousand lux, an effect due to the dark ceiling, and the worse if the flooring is of dark colour. This method of light control is of value in special situations, especially for certain processes of critical inspection (3.2).

**App. IV—2.5** *Modular luminaires for fluorescent tubular lamps.* Modular designs for luminaires have been developed to suit standard patterns of suspended ceilings. In some ranges of modular luminaires (sometimes also known as 'troffers') units matching the range but suitable for independent suspension may also be available (Sketch (a)), as well as those designed for mounting close-ceiling (Sketch (b)). Troffer luminaires may be designed as semi-recessed units (Sketch (c)) which have the advantage of

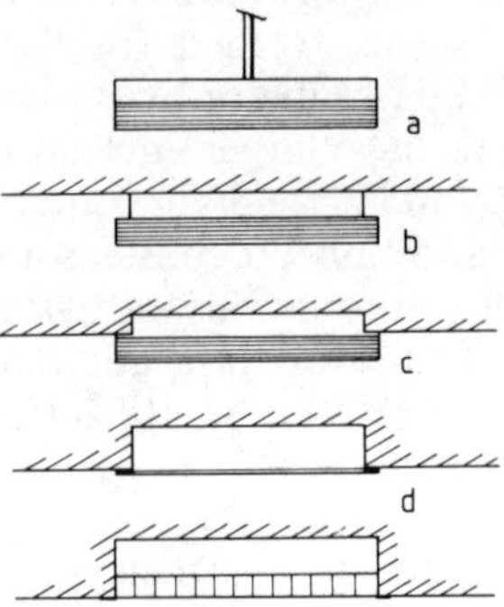


distributing a little light on to the surrounding ceiling, or at least creating a 'halo' of light which reduces the immediate brightness contrast between the luminaire and the ceiling. This advantage is not shared with fully-recessed troffer luminaires (Sketch (d)) which may be more fully mechanical integrated with the ceiling construction. For example, in some systems a grid of rails is erected to carry the luminaires and their wiring, the spaces between being infilled with lightweight panels to form a complete ceiling. The underside of such luminaires may carry diffusing panels, ribbed or prismatic clear plastic enclosures, or (as shown in Sketch (e)) may have a louvre panel placed below the lamps. Other luminaires of this kind may be fitted with low-brightness specular reflector units, with or without a base egg-crate louvre (see Appendix IV—2.4). Troffer units with louvred bases are used as *air-handling luminaires*, where air from the room is exhausted into a ducting system through the luminaires. If the air contains only dry dust particles, it can be arranged that the velocity of the exhaust air is too high to permit the dust to settle on the luminaire surfaces and lamp, and the cooling effect of the air current may enable the tubes to operate at a somewhat higher efficacy than tubes in still air.

**App. IV—2.6** *Coffered ceilings housing fluorescent tubular lamps.* One system of producing a coffered ceiling suitable for use in drawing offices has been discussed (20.2.1, Figure 20.2), and this idea can be applied in various designs suitable for office areas, showrooms and entrance halls. For example, instead of having a flat ceiling broken into cells by the small ceiling partitions, the ceiling itself may be formed in a series of shallow 'tents' of any convenient geometrical shape, all or some of which will be fitted with either a bare batten luminaire or an enclosed luminaire of some kind.

**App. IV—2.7** *Fluorescent lamp luminaires for hostile environments.* A variety of types of protected luminaires are employed in industrial lighting (Chapters 10, 11, 12) which are designed to withstand environmental conditions (cold, heat, dust, moisture, corrosion etc) and/or to be safe in

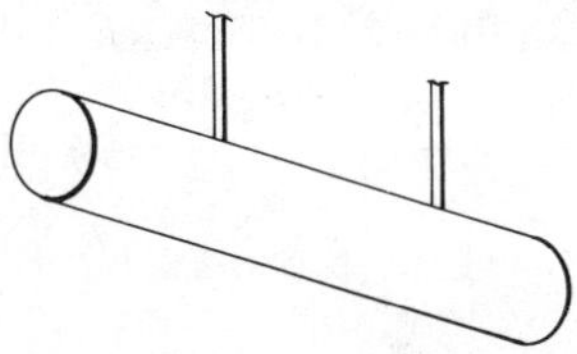

flammable atmospheres. One important type of protected luminaire is the pressurized luminaire (illustrated) which is discussed in 12.2.3. Some types of luminaires designed to withstand dust, damp or corrosion consist of a bare batten luminaire (App. IV—2.1) enclosed within a sealed translucent plastic enclosure.

## App. IV—3 Luminaires for HID lamps

**App. IV—3.1** *High-bay luminaires.* Luminaires for HID lamps (MBF, MBI and SON lamps) are made in dispersive and concentrating distributions for general lighting of industrial interiors. Control gear may be mounted local to the point of suspension, or, as shown in the sketch, a control-gear compartment may be located at the top of the luminaire. Originally, the term 'high bay' indicated that such luminaires could only be used at mounting

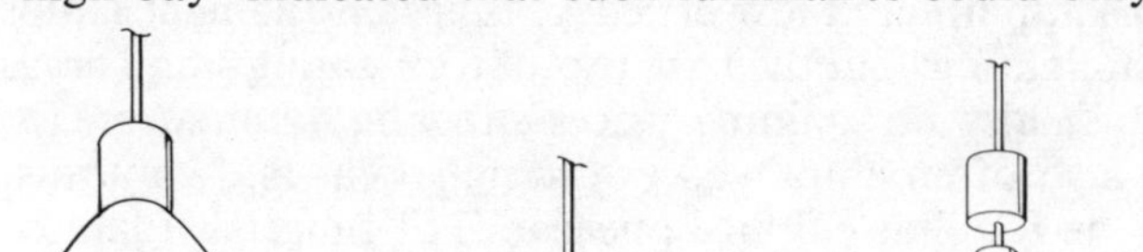


heights of 6 m or more above the working plane, but recent developments in these luminaires has produced versions which are suitable for mounting heights down to about 3.5 m. Reflector surfaces are commonly of anodized aluminium which has replaced the traditional vitreous enamel on steel which was employed for so many years. Reflectors may be through-vented (11.1.7) and may have 'overlamp removal of reflector', enabling the reflector to be removed for cleaning without removing the lamp. High-bay luminaires for hostile environments are made with suitable forms of protection against soiling (11.1.5) and to cope safely with hostile environments and flammable atmospheres. In the common form of high-bay luminaires, lamps have been mounted vertically, cap-up (a), but recently luminaires in which the lamp lies horizontally have been introduced, as shown diagrammatically in the accompanying sketch (b). This form of luminaire lends itself to non-symmetrical distributions which can be more efficiently applied to aisles and narrow buildings than luminaires with circular symmetrical distributions. For applications where the control of glare is not too critical, savings in installation cost may result from using a reflector-type HID lamps (MBFR, MBIR, SONR, preferably in a type of luminaire fitted with an 'anti-glare skirt' to reduce the brightness of the luminaire in normal directions of view (c). In practice it is found that reflector lamps in such luminaires return a very good service record in dirty atmospheres, for reasons explained in Figure 11.1.

**App. IV—3.2** *Angled/adjustable luminaires.* It should not be overlooked that reflectorized HID lamps are usually capable of being operated successfully at attitudes other than vertically mounted cap up, though not all patterns are 'universal burning'.

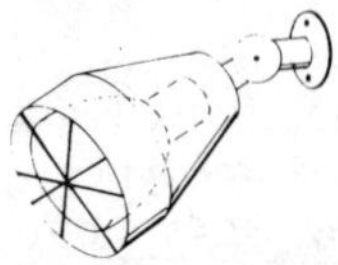

Where it is desirable to angle the flow of light, for example, in localized lighting and local lighting (2.2), a suitable luminaire to house a reflectorised HID lamp may be employed. If such units are placed within hand-reach height of a normal working position, then a pattern with a robust wire guard should be selected. These luminaires are these days preferred to traditional parabolic reflectors, and are more easily aligned for the particular purpose. They are also employed for cross-lighting to enhance vertical illuminance in tall buildings, or where the work entails the need for good illuminance on vertical planes at the workstation (Figure 8.1).

**App. IV—4** *Luminaires for tungsten-filament lamps.* Because of their disadvantages of short lamp life and low efficacy, tungsten-filament lamps are not used for general lighting; but they are useful for lighting small areas where there is no difficulty in gaining access to the luminaires for maintenance, and are also often employed in dining rooms and canteens, rest-rooms etc, where the familiar colour-appearance of tungsten-filament lighting is acceptable. Enclosed luminaires for one or two tungsten-filament lamps are often used for small passages, entrances and staircases; such units may be 'pilot lights' (normally left on at all times) and/or may be wired into the emergency lighting circuits as described in Chapter 6. For very small rooms, toilets and washrooms, units such as that shown in Sketch (a) may be mounted on the ceiling or upper walls, or an enclosed pendant unit, ceiling mounted, may be employed (b). Because small fluorescent lamp enclosed luminaires are now available (Appendix IV.5) these may be preferred for such applications on the grounds that they offer much longer lamp life coupled with smaller current consumption.

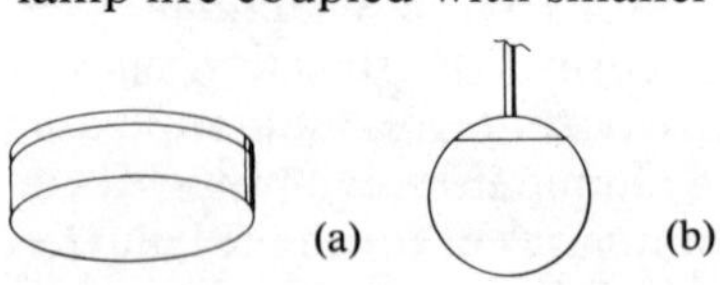


**App. IV—5** *Emergency lighting and pilot lighting luminaires.* Small enclosed luminaires for miniature fluorescent tubular lamps, viz 4 to 13 W, may be employed as lighting for small areas, as pilot lighting or as emergency lighting. The functions and applications of emergency lighting are discussed in Chapter 6. These units have largely suppressed the use of tungsten-filament lamp luminaires (Appendix IV—4.1) as units of the same appearance may be used for ordinary lighting, or for emergency lighting, the same unit sometimes being suitable both for outdoor and indoor use (Sketch (a)). Some are 'slave units', with one or more lamps fed from a central or zonal battery; others are equipped with their internal battery and charging

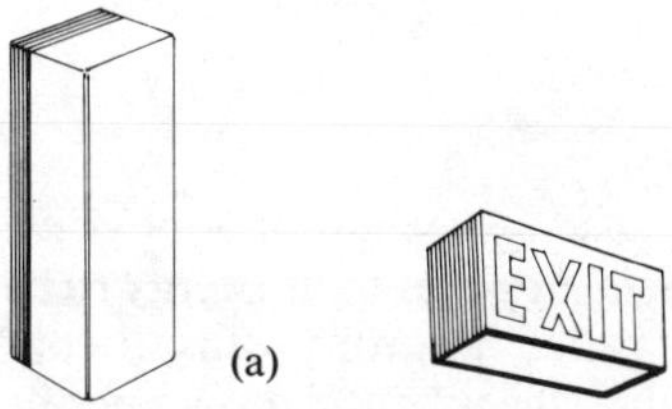


(a) (b)

circuit. Some patterns of Exit sign which comply with the technical requirements are designed to give lighting downward to illuminate the doorway or exit path which they mark (Sketch (b)).

**App. IV—6** *Handlamps.* Portable and mobile lighting equipment which is used within hand-reach height of a normal working position should be operated at reduced voltage (14.3, 14.4). Even with this important enhancement of safety, it is still desirable that handlamps and the like be robustly constructed and are earthed or 'double insulated'. The lamp should be enclosed in a cover of clear plastic, or should be guarded with a strong wire guard, and the design should be such that it is impossible to get at the lamp or

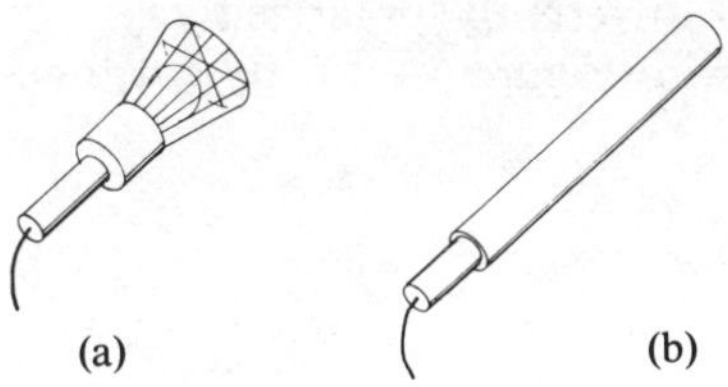

(a) (b)

the lampholder without the use of tools (a). Additionally, the design of handlamp should be suitable to withstand whatever environmental exist at the place of use. Note that in some industrial locations there may be technical reasons for not introducing an earth-lead into an otherwise earth-free location; in such a case, a 24 V handlamp should be used, supplied by a two-wire flexible cord. In locations where the environmental hazards are such that there is risk of the handlamp causing fire or explosion, a totally sealed lamp containing a fluorescent tubular lamp may be used (b). One type of handlamp of this kind cannot be relamped, and is disposed of when the lamp expires; others can be relamped by the makers and resealed, but should not be opened by the user.

Appendix V

# Polarized light

## App. V—1 Polarization phenomena

**App. V—1.1** Two alternative theories can be used to explain the physical behaviour of light: the *corpuscular theory* (which supposes light energy to be transmitted in discrete packages or quanta called photons, this theory enabling an explanation to be given for photo-chemical phenomena); and the *wave theory* (which supposes light energy to be transmitted in the form of continuous waves, this theory enabling explanations to be given of the phenomena of refraction and polarization). A ray of common light may be considered as vibrating in all directions across the direction of the ray, but in polarized light these vibrations occur in only one plane normal to the axis of the ray. Ordinary light may be considered as though composed of rays, each of which is a cylinder; polarized light may be considered as though composed of rays, each of which is a flat ribbon. This analogy permits the ready explanation of the phenomena of polarization and depolarization. Certain crystals (Icelandic Spar, for example) possess the property of internal double refraction, that is to say a ray of common light passing through one of these crystals is divided into two polarized rays which emerge from the crystal at slightly different angles, the plane of polarization of one ray being at right angles to that of the other. If these two differently polarized rays are recombined, common light is produced again (Figure V—1). By cutting a

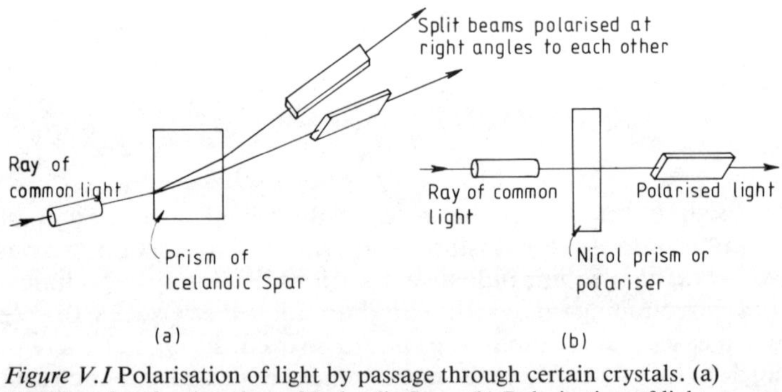


*Figure V.1* Polarisation of light by passage through certain crystals. (a) Double refraction in prism of Icelandic Spar. (b) Polarisation of light passing through a Nicol prism or polariser.

prism of Icelandic Spar and re-cementing the parts together in a particular way, it can be arranged that it transmits only one polarized ray; such is termed a *Nicol prism* or polarizer. Other crystals (e.g. tourmaline, iodo-sulphate of quinine or herapathite) have the inherent property of transmitting only one polarized ray. Such polarizers may be compared with a grating composed of narrow flat bars; through such a grating, a flat ribbon (polarized light) will pass readily; but a cylinder (common light) will not pass.

Similarly, the flat ribbon (polarized light) will only pass through the grating if its flat side is parallel to the grating (e.g. of same plane of polarization); thus, polarized light will pass unimpeded through a polarizer in the same plane of polarization, and will be completely stopped by a polarizer rotated through 90° along the axis of the ray. Ruled gratings which produce a polarizing effect are widely employed for photographic and instrument purposes.

**App. V—1.2** Light can also be polarized by reflection. The degree of polarization depends upon the angle at which the incident ray is reflected and on the number of times it is reflected. Maximum polarization in one reflection occurs at one particular angle; *Sir David Brewster's Law* states, 'The index of refraction is the tangent of the angle of polarization'. The maximum polarizing angle for water is 53°11′, and that for ordinary glass is 56°45′, while that for multiplate polarizers made of plastics materials is around 57°. At angles 20° greater or lesser than the polarizing angle, some eight reflections are required to achieve complete polarization. Multiplate polarizers (also known as pile-of-plate polarizers) (Figure V—2) are a means of polarizing common light, and may be incorporated into luminaires so that their light output is substantially polarized.

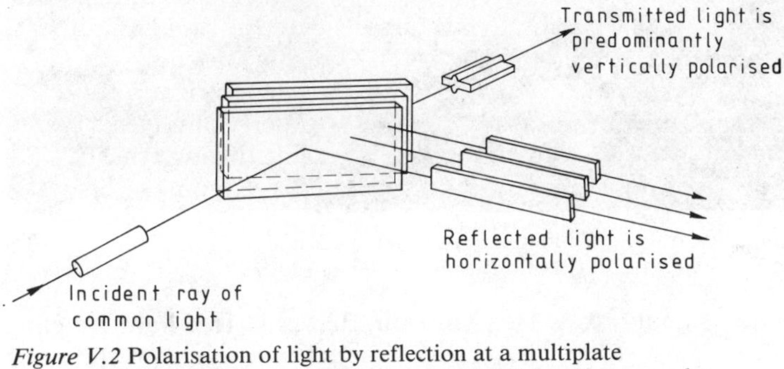


*Figure V.2* Polarisation of light by reflection at a multiplate polariser. Both the reflected and the transmitted light effects may be employed in the design of polarising luminaires.

## App. V—2 Applications of polarized light

**App. V—2.1** Polarized light is used in certain inspection processes (3.3.5a, 3.4.1h, 3.41i). It is sometimes applied in very critical colour-matching operations to standardize conditions and avoid apparent change of colour due to reflections (4.3.10). Of increasing importance is the use of polarized light to minimize reflections which otherwise impede vision; of these, two practical examples may be quoted: (a) When an operator is looking into a mass of clear liquid, e.g. water in a cannister cooling-pond in a nuclear power station, reflections from the surface of the water can be reduced greatly by the operator wearing polarizing spectacles; (2) troublesome reflections from the glasses over indicating instruments in control-boards can be minimized by illuminating the instruments with suitably polarized light, so that the brightness of the reflections is greatly reduced in the chosen angle of view.

**App. V—2.2** There is growing use of polarized light for general lighting systems where reflections may otherwise be troublesome. Tye[56] reports that polarized light is used in the Port of London Authority Navigation Service radar rooms as a means of reducing glare, and that it is also applied in

*Figure V.3* An application of localised luminaires emitting polarised light at Champion Spark Plugs Ltd. (Photo: Richard Daleman Ltd).

certain British Rail signal boxes. He points out that polarized light appears, under certain conditions, to be less bright than common light of the same illuminance, and makes claims of greater comfort for such installations. Tye also confirms the author's opinion that colour-rendering improvements result from the use of polarized light because the reflections from each particle of the coloured materials are not diluted by veilling reflections. This leads to the conclusion that suitable task visibility can be achieved at lower illuminance levels than are required with conventional lighting systems using common (unpolarized) light. While Tye quotes an impressive number of references, most of them originate in the USA, and it must be reported that experience in the use of polarizing materials in general lighting luminaires in the UK is at present limited. The author has noted that some enclosed refractor-diffuser luminaires emit partially polarized light, the polarization occurring by the effect of total internal reflections in the prisms formed on the surface of the diffuser; but in those examples he has seen, the polarization did not appear to be arranged to reduce glare. Enclosed refractor enclosures for fluorescent lamps can be designed with a multiplate polarization layer formed on the enclosing plastic material; light passing through is substantially polarized (Figure V—2), while that which is reflected back is further reflected before eventually passing through the polarizing layer. Thus, in such luminaires, the emitted light will all be substantially polarized, and, applied as local or localized lighting, they may add a measure of comfort in selected difficult visual tasks (Figure V—1).

# Appendix VI
# Units and conversion factors

LENGTH

| | | Metre | Foot | Inch |
|---|---|---|---|---|
| One metre | equals | 1 | 3.281 | 39.37 |
| One foot | equals | 0.305 | 1 | 12 |
| One inch | equals | 0.0254 | 0.0833 | 1 |

INCHES TO MILLIMETRES

| *Inches* | *0* | ¼ | ½ | ¾ |
|---|---|---|---|---|
| 0 | — | 6.35 | 12.7 | 19.05 |
| 1 | 25.4 | 31.75 | 38.1 | 44.45 |
| 2 | 50.8 | 57.15 | 63.5 | 69.85 |
| 3 | 76.2 | 82.55 | 88.9 | 95.25 |
| 4 | 101.6 | 107.95 | 114.3 | 120.65 |
| 5 | 127.0 | 133.35 | 139.7 | 146.05 |
| 6 | 152.4 | 158.75 | 165.1 | 171.45 |
| 7 | 177.8 | 184.15 | 190.5 | 196.85 |
| 8 | 203.2 | 209.55 | 215.9 | 222.25 |
| 9 | 228.6 | 234.95 | 241.3 | 247.65 |
| 10 | 254.0 | 260.35 | 266.7 | 273.05 |
| 11 | 279.4 | 285.75 | 292.1 | 298.45 |
| 12 | 304.8 | 311.15 | 317.5 | 323.85 |

ILLUMINANCE

| | *Lux* | *lm/ft² or foot-candle* |
|---|---|---|
| One lux equals | 1 | 0.093 |
| One lumen per square foot or foot-candle equals | 10.76 | 1 |

LUMINANCE

| | | *Apostilb* | *Foot-lambert* | *Lambert* | *cd/m²* | *cd/in²* | *Stilb* |
|---|---|---|---|---|---|---|---|
| One apostilb (lm/m²) | = | 1 | 0.0929 | 0.0001 | 0.318 | 0.000 205 | 0.000 0318 |
| One foot-lambert (lm/ft²) | = | 10.76 | 1 | 0.001 076 | 3.426 | 0.002 21 | 0.000 3426 |
| One lambert (lm/cm²) | = | 10 000 | 929 | 1 | 3183 | 2.054 | 0.318 |
| One candela per sq metre (cd/m²) | = | 3.14 | 0.292 | 0.000 314 | 1 | 0.000 645 | 0.0001 |
| One candela per sq inch (cd/in²) | = | 4869 | 452 | 0.487 | 1500 | 1 | 0.155 |
| One stilb (cd/cm²) | = | 31 416 | 2929 | 3.14 | 10 000 | 6.452 | 1 |

Multiply a unit in left hand column by the factor shown to convert to units along the top line.

Appendix VII

# Lightmeters

Illuminance is readily measured with a portable instrument called a *lightmeter* or luxmeter. The instrument is simply held in the plane of measurement and the illuminance in lux is read on the scale. Instruments of this kind commonly read from 10 lux to 5000 lux, but more sensitive patterns are available for reading the much lower levels used in emergency lighting and some outdoor lighting.

A lightmeter indicates the illuminance at the point of measurement only, not the average in the space. To find the average illuminance in an area at the time, it is necessary to divide the area into a number of equal areas which should be as nearly square as possible. The illuminance at the centre of each square is then measured, and the results averaged. The minimum number of equal areas required for accuracy can be determined by first working out the *room index* ($k$), thus

$$k = \frac{\text{Length} \times \text{Width}}{H_m \times (\text{Length} + \text{Width})}$$

where $H_m$ is the height of the luminaires above the plane of measurement. The working plane is usually taken to be 0.85 m (common bench-top height) unless the main plane of the work is known to be some other height above floor level. If the work is performed down to floor level, then the floor is taken as the working plane and plane of measurement.

The number of measurement points relates to the $k$ value thus: if $k$ is below 1, 4 points; between 1 and 2, 9 points; between 2 and 3, 16 points; and if 3 or above, 25 points. If the proposed points coincide with the luminaire positions, or are in constant relationship with the luminaire positions, increase the number of measurement points.

It is recommended that a suitable lightmeter be purchased and held available for use at every factory, for it is as essential as a thermometer and just as easy to use. In the UK it is usually possible to borrow a lightmeter from your Electricity Board (indeed, the sales engineer will probably be pleased to measure the illuminance for you), or from your lighting supplier or electrical installer. When the illuminance in an area has been measured (taking care to exclude daylight or subtract its reading from the total reading on the lightmeter), you will be able to compare the present illuminance with the recommendations of the *CIBS/IES Code*[5] appropriate to the type of work performed in the area.

The following are some British manufacturers of lightmeters:

Hagner International (UK) Ltd, 42 Little London, Chichester, Sussex, PO19 1PL.

Megatron Ltd, 165 Marlborough Road, London, N19.

Permic Emergency Lighting Ltd, PO Box 3, Chesterfield, Derbyshire, S40 1EX.

Salford Electrical Instruments Ltd, Peel Works, Barton Lane, Eccles, Manchester, M30 0HL.

Sangamo-Weston Ltd, Enfield, Middlesex.

Lightmeters are covered by BS 667:1968, *Portable Photelectric Photometers.* Those reading in the range 10-5000 lux full scale deflection are made to an accuracy of ± 15 per cent, though they may be used with an accuracy of ± 5 per cent if regularly calibrated and if the readings are weighted with suitable correction factors applied according to the calibration chart from a test house. Higher accuracy instruments are available.

Avoid subjecting the lightmeter to excessive vibration or knocks. Keep it in its case when not in use, following the maker's instructions as whether it should be stored or carried with the pointer locked or not. Before using the lightmeter, expose it for ten minutes to an illuminance that moves the pointer to about the centre of the scale; then subject it to several swings of up to full scale deflection by facing the cell to a suitable light source, but take care not to expose it to a greater illuminance than is catered for on the scale.

Some lightmeters have a scale-change switch. High resistance in the switch can cause inaccurate readings, so check one scale against another, viz, set the instrument where it records an illuminance which can be read on either of two scales, and compare the readings. With the cell covered, move the switch between its positions several times. Other instruments achieve scale change by placing masking multipliers on the cell. Some lightmeters have a separate light-cell which is connected by a flexible cable. The correct polarity is important. Do not shade the cell with your body when taking readings. For accuracy, repeat readings at the same point, but standing in different positions relative to the cell.

To take lighting readings in a room into which daylight penetrates, take the readings as described, then switch out the lighting and take the readings again. Subtract the second readings from the first set. Do not delay in this procedure, for daylight is very variable.

If the instrument does not read zero when no light is falling on the cell, it can be adjusted by the zero adjuster, a procedure that should rarely be necessary and which should be performed with care.

If the instrument is fitted with a pointer-lock, this can be used to take readings in difficult situations, thus: place the cell at the point of measurement, then lock the pointer. The reading can be read with the instrument in a convenient position.

A method of estimating the approximate reflection factor of a surface is to measure the illuminance at the surface ($E_1$), and then take a reading with the lightmeter cell facing the surface and held 300 mm from it ($E_2$). Then,

$$\frac{E_2}{E_1} = \text{Reflection Factor}$$

A slightly more accurate method is to place a sheet of white paper (at least

500 mm × 500 mm) over the surface, and take a reading of the lightmeter with the cell facing the paper and held 300 mm from it ($E_1$). The paper is then removed, and the reading taken again in the same fashion ($E_2$). For best results, the light should come to the surface reasonably normal to it and from a source of large area; also, the white paper used should be non-glossy. The measurement will be reasonably accurate on large areas of surface of uniform colour and non-glossy. Accuracy is improved if a black-lined tube is placed over the cell to prevent light from distant sources impinging upon it while carrying out the second estimating method described.

Appendix VIII

# Useful names and addresses

Here are listed organizations concerned directly or indirectly with the subject matter of this book.

The *British Approval Service for Electrical Equipment in Flammable Atmospheres* (BASEEFA), SMRE Laboratories, Harpur Hill, Buxton, Derbyshire.

The *British Electrical & Allied Manufacturers' Association Ltd* (BEAMA), Leicester House, 8 Leicester Street, London, WC2H 7BN.

The *British Standards Institution* (BSI), 2 Park Street, London, W1A 2BS.

The *Chartered Institution of Building Services* (CIBS), Delta House, 222 Balham High Road, London, SW12 9BS.

The *Commission Internationale de l'Eclairage* (CIE), Bureau Centrale de la CIE, 52 Boulevard Malesherbes, 75008 Paris, France. (The UK representative may be contacted through CIBS).

The *Electrical Contractors' Association* (ECA), Esca House, 34 Palace Court, London, W2 4HY.

The *Electricity Council for England & Wales* (EC), 30 Millbank, London, SW1P 4RD.

The *Health & Safety Executive* (HSE), Baynard House, Chepstow Place, London, W2 4TF.

*Her Majesty's Stationery Office* (HMSO), PO Box 569, London, SE1 9NH.

The *Illuminating Engineering Society* (IES). This society no longer exists as a separate entity, but is incorporated into the Chartered Institution of Building Services. However, some IES publications are still available from CIBS.

The *Lighting Industry Federation Ltd* (LIF), Swan House, 207 Balham High Road, London, SW17 7BQ.

The *National Inspection Council for Electrical Installation Contracting* (NICEIC), 237 Kennington Lane, London, SE11 5QJ.

The following publications carry information about lighting:

*Electrical Times* (weekly) and *Electrical Review* (weekly), published by IPC Business Press Ltd, Quadrant House, The Quadrant, Sutton, Surrey SM2 5AS.

*Lighting Equipment News* (monthly), published by Troupe Publications, Maclean Hunter Ltd, 30 Old Burlington Street, London W1X 2AE.

*Lighting Research & Technology* (quarterly) and
*Building Services* (monthly), both published by the Chartered Institution of Building Services.

Appendix IX

# Bibliography and Further Reading

(1) Lyons, S. L., *Exterior Lighting for Industry and Security,* Applied Science Publishers Ltd, Barking, Essex (1980).

(2) Carlton, Bill, *Industrial lighting must be tailored to real-life working conditions,* Electrical Times, Issue 4562 (14 March 1980).

(3) Health and Safety Executive, *Effective policies for health and safety,* HMSO (1980).

(4) Lyons, S. L., *Management Guide to Modern Industrial Lighting,* Applied Science Publishers Ltd, Barking, Essex (1972).

(5) Chartered Institution of Building Services, *IES Code for Interior Lighting,* (1977 Edition). Note: future editions will be titled 'CIBS Code for Interior Lighting'.

(6) Chartered Institution of Building Services, *Preprint of the National Lighting Conference at Canterbury* (1980). (Texts of 35 lighting papers presented at the Conference).

(7) *Interior Lighting Design,* 5th Edition, published by The Electricity Council and The Lighting Industry Federation Ltd, ISBN 0 901986 10 0, (1977).

(8) The Electricity Council, *Daylighting of farm buildings,* Technical Information AGR 5-6, (1977).

(9) BS CP3: Chapter 1: Part 1: 1964, *Daylighting.*

(10) The Electricity Council, *Essentials of Good Lighting,* (1974) (Out of print).

(11) Bellchambers, H. E., and Phillipson, S. M., *Lighting for Inspection,* IES Trans., Vol. 27, No. 2 (1962).

(12) Lyons, S. L., *The Influence of Lighting on Industrial and Domestic Accidents,* Journal of the Junior Institution of Engineers, Vol. 59, Part 9, (1949).

(13) *Occupational Safety and Health,* published monthly by the Royal Society for the Prevention of Accidents.

(14) *Annual Reports of HM Chief Inspector of Factories,* Health and Safety Executive, published by HMSO.

(15) *Regulations for the Electrical Equipment of Buildings* (15th Edition) (1981), Institution of Electrical Engineers, London. (Also familiarly known as the 'Wiring Regulations'.)

(16) *Lighting in Industry* (3rd Edition 1967), published jointly by The Electricity Council and the (former) British Lighting Council. (Out of print).

(17) *Colour and lighting in factories and offices,* published by the British Colour Council, London, (1956).

(18) BS 5252:1976 *Framework for colour co-ordination for building purposes.*

(19) *British Standards Yearbook,* published by British Standards Institution, annually.

(20) BS 1710:1975, *Identification of pipelines (*gives colour specifications to BS4800).

(21) Publication 13.2, *Colour rendering of light sources,* Commission Internationale de l'Eclairage.

(22) Lockyer, K. G., *Introduction to Critical Path Analysis,* 3rd Ed, 1969, Pitman.

(23) *Protection of eyes Regulations 1974 No 1681, as amended 1975 No 303,* published by HMSO (01 104 168 13-ISBN and 01 1050303 1-ISBN).

(24) BS 5266:Part 1:1975 (with amendment AMD 3112), *Emergency lighting, Part 1. Code of practice for the emergency lighting of premises other than cinemas and certain other specified premises used for entertainment.*

(25) *Fire Precautions Act,* 1971

(26) *Health & Safety at Work Etc Act,* 1974

(27) BS 4533:--: *Electric luminaires*

(28) *Integrated Design,* The Electricity Council, EC/E/I/2685, 1969.

(29) Hopkinson, R. G., and Kay, J. D., *The Lighting of Buildings,* Faber & Faber, 1969.

(30) *Heat Recovery,* EC3472R, The Electricity Council, 1979.

(31) *Heat Pumps, the energy savers,* EC3839, The Electricity Council, 1979.

(32) *Comfort; what it is and how it affects your work,* EC3437, The Electricity Council, 1978.

(33) Lyons, S. L., *Electrical Times,* 4/11th April 1980, p.12.

(34) BS 5394:Part 1:1976, *Specification for radio interference limits and measurements for lighting equipment,* Part 1: *Luminaires for tubular fluorescent lamps.*

(35) Bowtell, J. N., *Suppression of Radio Interference from fluorescent lighting equipment,* GEC Journal of Science and Technology, Vol.38, No.1, 1971.

(36) *Harmonics and waveform distortion due to lighting,* Pubn.G5/2, available from Distribution Engineering Section, The Electricity Council.

(37) *Daytime lighting in buildings,* Technical Report No 4, with supplement, IES/Chartered Institution of Building Services, London, 1972.

(38) BS 5345: - - *Code of Practice for the selection, installation and maintenance of electrical apparatus for use in potentially explosive atmospheres (other than mining applications or explosive processing and manufacture).*

(39) CP 1003: - -, *Electrical apparatus and associated equipment for use in explosive atmospheres of gas or vapour other than mining applications.*

(40) IES Technical Report No 9, *Depreciation and Maintenance of Interior Lighting,* 1967, published by the former Illuminating Engineering Society (now incorporated into the Chartered Institution of Building Services).

(41) BS 4343:1968, *Industrial plugs, socket-outlets and couplers for a.c. and d.c. supplies.*

(42) *IES Lighting Guide No 3: Building and civil engineering sites,* 1975, now published by Chartered Institution of Building Services.

(43) BS 4363:1968, *Distribution units for electricity supplies for construction and building sites.*

(44) BS 5489: -.-. (in 9-parts); *Code of Practice for roadlighting.*

(45) Baker, J. E., and Lyons, S. L., *Lighting for the security of premises,* Lighting Research & Technology, Vol. 10, No.1, 1978.

(46) Keward B. A., Ogden K. W., and Parker P.D., *'Compatibility of new discharge sources',* IES/CIBS National Lighting Conference, 1978.

(47) Publication No 29, *Guide to interior lighting,* Commission Internationale de l'Éclairage.

(48) *The Food Hygiene (General) Regulations,* SI No 1601, 1969, HMSO.

(49) Lyons, S. L., *Lighting in the Clothing Industry,* Light & Lighting and Environmental Design, October 1971.

(50) Dipl-Ing. Carl-Heinz Herbst, Wiesbaden, *The Effect of Light upon Workers,* Electrizität, No 11, 1968, pp.294-300.

(51) Pritchard, D. C., *Environmental Physics: Lighting,* Longmans, 1969.

(52) BS 1853: -.-, *Tubular fluorescent lamps for general lighting service;* Part 1:1979 *Specification for internationally specified lamps;* Part 2:1979 *Specification for lamps used in the United Kingdom not included in Part 1.*

(53) BS 4211:1967, *Steel ladders for permanent access.*

(54) *Better industrial lighting—the economic case,* published by the former British Lighting Council, 1966, (UDC:628.9:725:4).

(55) *Office Planner,* A4 Publications Ltd (Benn Brothers Ltd), 1976, ISBN 0 510 07017-5.

(56) R. L. M. Tye, *Polarisation helps reduce interior lighting costs,* Electrical Review, Vol.203, No 10, 15 September 1978.

Appendix X

# Buyer's Guide to products of UK lighting manufacturers

The following list of British lighting manufacturers includes many who are Members of the Lighting Industry Federation Limited, and these are marked thus*. The data in the list was up to date at the time of compilation (September 1980). Some products of particular manufacturers may not be indicated; some data could not be obtained in time for inclusion.

The letter symbols following each entry signify as follows:
A—Tubular fluorescent lamps; B—HID lamps; C—Control gear for fluorescent tubular lamps; D—Control gear for HID lamps; E—Adjustable task lights; F—Industrial filament-lamp luminaires; G—Commerical luminaires for fluorescent tubular lamps; H—Industrial luminaires for fluorescent tubular lamps; I—Commercial luminaires for HID lamps; J—Industrial luminaires for HID lamps; K—Luminaires for Zone 1; L—Luminaires for Zone 2; M—Emergency lighting; N—Photoelectric controls; O—Polarised light equipment; P—Plastic louvres etc for light control.

Alex Eng Co Ltd,* Longmead, Shaftsbury, Dorset, SP7 8PL, F G H I M

Allom Lighting Ltd,* Coombe House Annexe, St George's Sq, New Malden, Surrey, KT3 4HZ. E G H I M

Alumex Lighting (Northern) Ltd,* 1/5 Tannoch Drive, West Lenziemill Industrial Estate, Cumbernauld, G67 2SX. C D E F G H I J K L M

Anglepoise Lighting Ltd,* Unit 51, Enfield Industrial Estate, Redditch, B97 6DR. E F G H.

Arrow Plastics Ltd,* Arrow Works, Hampden Road, Kingston-upon-Thames, Surrey, KT1 3HQ. P

Ascog Ltd,* Nathan Way, Woolwich/Thamesmead, London, SE28 0AZ. C D E F G H I J

Ascot Lamps & Lighting Ltd,* Cray Avenue, St Mary Cray, Orpington, kent, BR5 3PS. Orpington 36231. A G

B B I Lighting Ltd,* Rankine Road, Daneshill Estate, Basingstoke, Hants, RG24 0PH C D E F G H I J K L M

Balena Lighting Productions Ltd,* Cabot Lane, Creekmoor, Poole, Dorset, BH17 7BY. E F G M

Beta Lighting Ltd,* 383-387 Leeds Road, Bradford BD3 9LZ. E F G I M

British Electric Lamps Ltd,* Spencer Hill Road, Wimbledon, London, SW19 4EL. A

Ceag Ltd,* PO Box 10, Langdale Road, Barnsley, South Yorkshire (0226 6842) M

Concord Lighting (International) Ltd,* Rotaflex House, 241 City Road, London, EC1P 1ET. D E F G I

Control Components Ltd,* Princewood Road, Corby, Northamptonshire, NN17 2AP. C D

Courtney, Pope Lighting Ltd,* Amhurst Park Works, Tottenham, London, N15 6RB. C D G H I J L M

Crompton Parkinson Ltd, 50/52 Marefair, Northampton, NN1 1NY. A B C D F G H I J K L M

Cryselco Ltd,* Kempston Works, Woburn Road, Kempston, Bedford MK42 7QB. A B C D F G H I J

Daleman Ltd, Richard, 325 Latimer Road, London W10 6RE (01-969 7455) O

Designplan Lighting Ltd,* 18 Grove Road, Sutton, Surrey, SM1 1BW. H L M

Edison Halo Ltd,* Eskdale Road, Uxbridge Industrial Estate, Uxbridge, Middlesex, UB8 2RT. E I M

Electronic & Fluorescent Accessories Ltd,* EFA House, 808-809 Oxford Avenue, Slough, SL1 4LN. C D

Energy Conservation Systems Ltd, Gresham House, Twickenham Road, Feltham, Middlesex, TW13 6HA. N

Ensel Electric Co Ltd,* Emily Place, Queensland Road, London, N7 7DQ C G H

FitzGerald Lighting Ltd,* Normandy Way, Walker Lines, Bodmin, Cornwall. C G H K M

Fluorescent Controls & Lighting Ltd,* Millfield Industrial Estate, Bentley, Doncaster, Yorkshire. C D F G H I J M

Heyes of Wigan,* PO Box 60, Walthew House Lane, Off Martland Mill Lane, Wigan, WN5 0LD. F H J K L

Holophane Europe Ltd,* Bond Avenue, Bletchley, Milton Keynes, MK1 1JG. F G H I J K L

Horsell Electrics Ltd,* 129-135 Beulah Road, Thornton Heath, Surrey, CR4 8XX. C G H I M

Hume-Atkins (Lighting) Ltd,* PO Box 19, Langdale Road, Barnsley, South Yorkshire, S71 1DT. E G H I M

JSB Electrical Ltd, Manor Lane, Holmes Chapel, Crewe, Cheshire, CW4 8AB. C D M

K-S-H Plastics (UK) Ltd, Units 2 & 3, The Old Airfield, Warboys, Huntingdon, Cambridgeshire PE17 2SH. P

Lab-Craft Ltd,* Church Road, Harold Wood, Romford, Essex, RM3 0HT. M

Linolite Ltd,* Pier Road, Feltham, Middx, TW14 0TW. F G H M

Lumitron Ltd,* Chandos Road, London, NW10 6PA. C D E F J H I J M

Luxram Electric Ltd,* 70/72 Great Eastern Street, London, EC2A 3JS. A B E G

M & P Fluorescent Fittings Ltd,* Bridge Road, Haywards Heath, Sussex, RH16 1UA. A B C D E G H I J M

Marlin Lighting Ltd,* (Merchant Adventurers Ltd), Hanworth Trading Estate, Hampton Road West, Feltham, Middlesex, TW13 6DR. E G H I M

Menvier (Electronic Engineers) Ltd,* Southam Road, Banbury, Oxfordshire, OX16 7RX. M

Moorlite Electrical Ltd,* Burlington Street, Ashton-under-Lyne, Lancs. OL7 0AX. A B C D E F G H I J M

S J Morley Ltd,* 329 Bowes Road, London, N11 1BA. C D G H I J M

Omega Lampworks Ltd,* Albany House, Burlington Road, New Malden, Surrey, KT3 4NJ. A B C M

Osram (GEC) Ltd,* PO Box 17, East Lane, Wembley, Middlesex, HA9 7PG. A B C D F G H I J K L M N

W J Parry & Co (Nottingham) Ltd,* Victoria Mills, Draycott, Derbyshire, DE7 3PW. C D

Philips Lighting,* PO Box 130, 17 Beddington Farm Road, Croydon, CR9 4JN. A B C D E F G H I J K L M N

Phosco Ltd,* Lower Road, Great Amwell, Ware, Hertfordshire, SG12 9TA. J

Pope's Electric Lamp Co Ltd,* Cray Avenue, St Mary Cray, Orpington, Kent, BR5 3PS. A B

Poselco Ltd,* Walmgate Road, Perivale, Greenford, Middlesex, UB6 7LX. C D G H I J M

RADA + SLR,* Hollies Way, High Street, Potters Bar, Hertfordshire, EN6 5BH. E F G H I J M

Security Lighting Ltd, 27 Breakfield, Ullswater Crescent, Coulsdon, Surrey, CR3 2HS. M

Simplex-GE Lighting,* Groveland Road, Tipton, West Midlands, DY4 7XB. A B C D F G H I J K L M

Stella Lamp Co Ltd,* 33/34 High Street, South Norwood, London, SE25 6HF. A B C D F G H I J K M

Tamworth Electrical Eng Co Ltd,* Harcourt, Halesfield, Telford, Shropshire. C F G H M

Thorn Lighting Ltd,* Thorn House, Upper St Martin's Lane, London, WC2H 9ED. A B C D E F G H I J K L M N

F W Thorpe Ltd,* Facet Road, Kings Norton, Birmingham, B38 9PU. F G H I J L M

Thousand & One Lamps Ltd,* 108 Bromley Road, London SE6 2UX. E F G H

Transtar Ltd,* Prince Consort Road, Hebburn, Tyne & Wear, NE31 1DU. C D

Transtrip Ltd,* Bridge Road, Cirencester, Gloucestershire, GL7 1NQ. M

Troughton Young,* Station Road, Gerrards Cross, Buckinghamshire, SL9 8EN. A B C D G H I J

Turnock Ltd, George,* Navigation Street, Walsall, West Midlands, WS2 9LU. E F

Victor Products (Wallsend) Ltd,* Norham Road, West Chirton Industrial Estate, North Shields, Tyne & Wear, NE29 7XW. F H J K M

D Walter & Co Ltd,* Kangley Bridge Road, Lower Sydenham, London, SE26 5AP. E G H I J M

Westair Dynamics Ltd,* Thames Works, Central Avenue, East Molesey, Surrey, KT8 0QZ. C D K L

# Index

The references given are to the first mention or the most important mention of the indexed item in the text.